Sentinels of the South Atlantic & Gulf Coasts

Dale Manuel

Published by the Coast Defense Study Group Press
2026

Cover Photographs:

Front cover:

Fort Morgan, Alabama, photograph by John Weaver

Rear cover

Upper Photo: The Advanced Redoubt of Fort Barrancas, Florida
Middle Photo: The parade of Fort Pulaski, Georgia
Lower Photo: the sallyport of Fort Morgan, Alabama
Photos by Mark Berhow

Sentinels of the South Atlantic & Gulf Coasts

Dale Manuel

Published by the Coast Defense Study Group Press
2026

PLEASE DIRECT ANY COMMENTS OR CORRECTIONS TO THE PUBLISHER - INFO@CDSG.ORG

IBSN 979-8-9947989-0-4 (Paperback B&W)

Library of Congress Control Number: 2026938648

Library of Congress Cataloging-in-Publication data
Title / Author
p. cm.
Includes bibliographical references and index.
Library of Congress Control Number: 000000000
ISBN 979-8-9947989-0-4 (p.b.)
1. Military History, 2. Artillery I. Dale Manuel

First Edition: June 2026
Printed in the USA by Ingram Spark

Table of Contents

Photo of Dale Manuel, complete with cowboy hat Dale Manuel Collection

Dale Manuel has always had a keen interest in the history of the gulf coast of the United States. He is the author of many articles and several books on the military forts that have dotted the coastlines of Texas, Louisiana, Missisippi, Alabama, Florida, Georgia, South Caroline, North Carolina, Virginia and up into New England. His five book series on the forts of Norfolk, Georgia, Florida, Mobile Bay and Lousiana have now long been out of print.

The Coast Defense Study Group recognizes the importance of Dale's work and has sponsored this second edition re-publication of his books in this single volume.

This book is a compilation of five volumes written by Dale covering the key areas of the southern Atlantic and Gulf coasts of the United States. It covers all periods of coast-defense history and contains an incredible number of illustrations to bring coast defense to life. This extensively revised edition combines the content of the five books, and edited by John Weaver, who also updated the current status of a number of these forts as of 2025. The illustrations were prepared by Mark Berhow and John Weaver, with several of the illustrations originally published by Manuel replaced by higher-resolution photographs from the Berhow and Weaver collections. Many of the illustrations have been enhanced to increase the resolution and contrast with software programs. The reader should be aware that this process can add/edit some of the finer details in the illustrations.

Fort Morgan gorge, photograph from outworks Photo by Mark Berhow

Introduction to the Systems of American Coastal Defense

Colonial Defenses

The Treaty of Paris, signed in 1783 thus ending the Revolutionary War, created the United States of America. With the signing of this treaty, all colonial forts in the new country became the property of the USA – the first defenses of our new nation. An example of this was Fort William and Mary in Portsmouth, New Hampshire, that was renamed Fort Constitution and manned by New Hampshire militia. Likewise, British colonial Fort Groton, protecting the Thames River in Connecticut, was renamed Fort Griswold when it became an American fort.

The First System of Coastal Defense

In 1794, war clouds loomed over this land. The United States was being drawn into the war that was raging between England and France. Under this threat, President George Washington initiated construction of a coastal defense system along America's Maritime Frontier known as the "First System" of American Coastal Defense (1794-1804). This system utilized open batteries with earthen ramparts that were located on strategic points around major harbors.

The Second System of Coastal Defense

In 1807, the British frigate Leopard attacked the American frigate Chesapeake off Virginia. This provocation caused President Jefferson and Congress to initiate the "Second System" (1807-1814) of defenses. One million dollars was appropriated, and Major Joseph Swift of the elite Army Corps of Engineers was to supervise construction. During the era of the "Second System" the all-masonry fort and the casemated gun emplacement were developed. In order to achieve a wide and continuous angle of fire, semicircular fronts were the most widely used design. Many forts had bastions and demibastions for landward defense.

During the War of 1812, Washington D.C. was captured and burned by the British. American General Andrew Jackson had many difficulties defending New Orleans and Mobile. Small Fort St. Philip, though, kept a British fleet from reaching New Orleans. In the same war, Fort McHenry held a British fleet at bay in Baltimore Harbor. Although these were successful defenses by forts, the inadequacies of the first two systems were apparent. A more comprehensive system of permanent fortifications was needed.

Fort Norfolk from the air. Army Corps of Engineers

An 1853 View of Craney Island. Harpers Weekly

The Third System of Coastal Defense

In 1816, a board of officers met to create a comprehensive, integrated, and permanent system of coastal defense. The system they initiated was originally called the Permanent System, but later became known as the "Third System" (1816-1867) of Coastal Defense. It was a strategic look at coastal defense, including the navy, communications, militia artillery, and masonry fortifications. These masonry forts and appurtenant structures had overwhelming firepower and flanking defenses. By 1822, a survey had been made of our coastline, both the Atlantic Coast and the coast of the Gulf of Mexico (now Gulf of America). From this survey the board made its recommendations for the locations of navy yards and fortifications, as well as improvements in communications, such as the Second National Road.

Two members of the board, Simon Bernard and Joseph Totten, would emerge to be the most influential engineers in the history of the System. The board's first leader was American engineer Joseph Swift, but he was soon supplanted by French military engineer and former aide-de-camp to Napoleon I, General Simon Bernard. Bernard's designs were greatly influenced by the grandiose style of French engineer Vauban. When Bernard resigned to return to France in 1831, Joseph Totten became the chief designer of the System's forts. New England born, Totten was perhaps the finest military engineer America has ever produced. Totten's innovations included a reduction in the size of embrasures and the development of iron embrasures with shutters.

The American system was greatly influenced by the French school of military architecture. The most widely used designs of this system had irregular and polygonal forms with bastions or counterscarp galleries. Most were casemated and had multiple tiers with the top tier mounting guns en barbette. These forts were designed in an age of slow, deep-draft sailing ships that were wind dependent, an age of less-than-accurate and less-destructive smoothbore cannon.

The golden age of fort building along our coast lasted about fifty years, the longest peacetime effort at national defense in our history. In the east, a line of forts would span the coast from Grand Terre Island, Louisiana, to Bucksport, Maine. Forts were built on Lake Champlain and the Great Lakes. The two defenses built on San Francisco Bay were the only forts built on the West Coast, but an independent masonry battery was built on the Columbia River and multiple supporting batteries were built around San Francisco Bay.

At the beginning of the Civil War, both North and South put great stock in these forts to deliver victory. Because the Civil War was the first modern war, it had many technological innovations such as ironclad steamships of war and rifled guns.

The Modern Defenses of the Endicott Period (1890-1907)

In 1885, Congress became alarmed by the growth of European naval powers and the deteriorating condition of U.S. coastal defenses. In that year, Secretary of War William Endicott convened a board to study the latest armament technology and fortification types. The board's recommendations were the basis for those U.S. coastal defenses built during what was eventually known as the Endicott Period.

The Endicott System was a continually evolving, modern system that coordinated breechloading rifled steel cannon, rifled steel mortars, rapid-fire weapons, electrically-controlled mine fields, and surveillance systems to guard the major harbors. The period was known for its concrete batteries. Each battery and gun was designed for a specific defensive purpose and utilized overlapping fields of fire and indirect fire control. Not only would the enemy fleet have to deal with artillery, but it would have to maneuver through an electrically controlled submarine mine field.

The system's weapons development began in 1888. Battery construction began in the early 1890s, intensified during the Spanish-American War of 1898, and continued into the early twentieth century. Batteries contained external structures such as crow's nests, gun platforms, and Battery Commander's stations and internal rooms such as magazines, storerooms, plotting rooms, and ammunition-service facilities.

The long-range, steel breechloading rifles were in various calibers of 6-, 8-, 10-, and 12-inch monsters on "disappearing" and barbette carriages. A 12-inch rifle could fire a 1,070 lb. shell approximately eight miles. The big rifles, mounted on "disappearing" carriages, were loaded behind the protection of the concrete parapet and then elevated over it. Upon firing, the recoil returned the gun to the rear and downward into its loading position. This action concealed the gun and provided safety for the gun crews. One reporter described the window-rattling firing of the big guns as follows: "With a clang, the breech is closed and slowly the big gun rises with its muzzle pointing out to sea. With a blinding flash and deafening roar, the projectile is on its way. Its passage is marked by a loud rumbling noise ending with a thud as it strikes the water."

Heavy 12-inch rifled mortars were mounted in pits of concrete and earth. Mortars had a range between 1.25 and 8.6 miles. Range depended on the model and weight of the projectile. Projectiles weighed between 700 and 1,046 pounds. The powder charges varied with the zones and ranged from 48 to 65 pounds. The mortars were normally mounted in groups of four per pit. When fired, the batteries concentrated a shotgun-like pattern of high trajectory, plunging fire that would have been effective in piercing the decks of enemy ships. So great was the concussion from firing the mortars in groups that a warning to households preceded practice firing. All windows were raised, and China was removed from shelves.

Rapid-fire rifles were in calibers of 3-, 4-, 4.7-, 5-, and 6-inch on various masking parapet, pedestal, and pillar mounts. A 3-inch rifle could fire up to ten rounds per minute. These flat trajectory weapons guarded the submarine mine fields from fast torpedo boats and minesweepers.

Mines, originally called torpedoes, were an important element of this defense package. A mine field could be laid in the main channel in time of war and remain harmless until armed electrically from a mining casemate on shore. Cables linked the submarine mines to the mining casemate. When not deployed, the unloaded mines were stored in a torpedo (mine) storehouse. Mine cables were stored on reels in water tanks to prevent rusting. The cables were also tested for electrical integrity in these facilities.

When the Taft board convened in 1905, technical changes were recommended for the Endicott batteries and the newer Taft-era (1907- 1920) batteries. Some of the changes were electrification, communications, ammunition service systems, and fire control techniques. Fire control techniques advanced over time, allowing the guns to be fired with pinpoint accuracy. Most batteries received power-generating plants to charge storage batteries and provide power for ammunition hoists, gun control motors, telephone systems, and general lighting.

In WW I, many guns and mortars were removed to serve as railroad artillery in France. Because of modern developments in armament, modifications to coastal defenses were made. The Army favored mobile artillery over fixed armament.

Construction of the Fortifications

The three main objectives of the Permanent or Third System were permanence of structure, overwhelming firepower, and landward defense. Permanence of structure was achieved by the use of masonry. Generally, in the North, granite was used, and in the South, brick was used due to ease of access to those materials. Overwhelming firepower was achieved through the use of multiple tiers of guns and polygonal form. Landward defense was achieved by a series of obstructions and flanking structures to protect the walls from direct fire and from escalade.

The first land obstacle was an earthen glacis which created a protective rise on the plane of the fort. The next obstacle was a low wall, or revetment, that marked the covered way. After the covered way came the wet or dry ditch. Then came the scarp wall of the fort. To prevent escalade of the scarp, flanking structures such as caponiers and bastions, and reverse-fire counterscarp galleries were used.

The size and form of forts was normally determined by the location and size of the site on which they were to be constructed and the size of the local population from which militias would be drawn. Usually, the larger the land front, the larger the fort. Regular polygonal forms were used to give continuous lines of fire on all sides.

Multiple tiers were used to maximize and concentrate firepower. In order to build the multiple tiers, a system of barrel-vaulted casemates and galleries was devised to support the tremendous weight of the iron guns. Each arch of the vaulting would equally distribute the weight to support walls and columns. In the vaulted casemates, the scarp wall narrowed to about five feet thick for the gun embrasures. The gun carriages were attached to the embrasures by iron tongues and iron pintles. To position the gun, it was rolled on an iron and granite traverse arc.

Reverse arches were an engineering design technique that partnered with the overhead arch to distribute weight evenly on wet soil and sand. This arrangement balanced the stresses on the foundation caused by the masonry and the massive cannon.

In order to remove the choking, sulfuric smoke of ignited black powder from the casemates, vents were built into the scarp walls and casemate arches. The vents were located near gun emplacements. Also to remove smoke, the vaulted galleries often opened out to the parade.

A magazine was designed to keep the fort's supply of black powder safe and dry. Wood lining and brass hinges for the doors prevented sparks. The magazines were well ventilated to keep them dry. Periodically, the barrels of powder were removed from the magazines and rolled to prevent caking.

Most of the "Second" and "Third" system forts had hot-shot furnaces built within their walls or in external batteries. The hot-shot furnace was used to heat shot red hot. Hot-shot could be fired at ships where it could start fires or ignite the magazine.

Totten embrasures and shutters offered guns crews more protection. The shutters were closed while loading the casemated guns. The shutters would be opened when the gun was rolled forward in order to fire it through the embrasure.

The engineers added many features to fort construction that were throwbacks to medieval castles. Wet moats, drawbridges, portcullis grates, machicoulis overhangs to fire down on an enemy, musket embrasures in the sallyports used to fire on trapped victims like the murder holes of castles, and even citadels that resembled keeps were all ancient ideas.

The Chesapeake-Leopard Affair

On January 22, 1807, the frigate USS *Chesapeake* set sail from Gosport Navy Yard carrying Commodore James Barron, who was to assume command of the Mediterranean fleet. Just off the Capes of Virginia the American frigate was intercepted by the British frigate HMS *Leopard* under the command of Captain S.P. Humphreys, who demanded to search the ship for British deserters. When Barron refused to allow the search, the *Leopard* opened fire. Not one of the *Chesapeake's* 36 guns were loaded while all the *Leopard's* 56 guns were ready for action. The *Chesapeake* only got off one shot when a burning coal from the ship's galley was carried aloft to fire only one gun. Four of the *Chesapeake's* crew were killed and twenty were wounded in the exchange.

A British boarding party seized four crewmen. One was later hanged as a deserter. After four years of diplomatic protests, the U.S. secured the release of two of the surviving sailors, the third having died while in British service. Commodore Barron was court-martialed and found guilty of negligence. He was suspended from the navy for five years. News of the incident provoked a furious outcry throughout the U.S. In Hampton, mobs broke casks of water that had been contracted by the British Navy. Barron would later kill Stephen Decatur in a duel because he had sat on his court-martial.

On July 2, 1807, President Thomas Jefferson expressed his indignation over the affair by ordering all British naval vessels to leave American ports. On December 22, 1807, the Embargo Act was passed.

The Act prohibited American ships from trading with any foreign country. The British economy was scarcely affected, but the American economy was nearly ruined. The Second System of coastal defenses was established in 1807. Before the establishment of the Second System, the entire Maritime Frontier of the United States was nearly undefended.

The affair would not be the only incident of the death and impressment of U.S. sailors as the British policies would continue. The affair, along with the continued impressment of our sailors, led to a declaration of war in 1812.

The *Chesapeake's* hard luck would continue during the War of 1812. On June 1, 1813, Captain James Lawrence boldly attacked the HMS *Shannon* with a green crew. After a broadside from the *Shannon* the British swarmed aboard. Lawrence was shot through the lungs and gasped the famous words "Don't Give Up the Ship" before he died.

Drawing of the battle between the USS Chesapeake and the HMS Shannon
Dale Manuel Collection

James Lawrence Dale Manuel Collection

After a great loss of life, the *Chesapeake* was forced to surrender to the British.

In 1819, the *Chesapeake* was sold for salvage. Timber from the gundeck was used to build a water mill in the village of Whickham in Hampshire, England.

The Shannon taking the Chesapeake into Halifax Dale Manuel Collection

Only one other navy vessel was named *Chesapeake,* and that one for only a short time. A Naval Academy training ship that was launched in 1899 as the *Chesapeake* was renamed the *Severn* by the navy. It seems that the navy was wary of the inglorious reputation of the name *Chesapeake.*

The First and Second System Defenses of Hampton Roads

Fort Nelson, Elizabeth River Begun 1794

In 1794, war threatened again. The United States was being drawn into a war between England and France. Under this threat, President George Washington initiated the First System of coastal defense. In 1794, irregularly traced forts were built 1,000 yards diagonally apart on both sides of the Elizabeth River to protect the harbor of Norfolk and the Navy Yard. The fort on the east side was Fort Norfolk.

The fort on the Portsmouth side was Fort Nelson. Together these forts could create a crossfire on any enemy vessels that would attempt to ascend the river. The 1794 fort replaced an earlier fort of the Revolutionary War that had been captured and blown up by the British. It was named for the Revolutionary War leader General Thomas Nelson, who was later Governor of Virginia.

In 1794, Fort Nelson was designed by the Swiss engineer John Jacob Ulrick Rivardi. The fort consisted of an enclosure, traverses, and batteries behind an earthen rampart with embrasures through which the cannon were fired. The revetment and embrasures were framed by wooden joists and faced with planks to hold back the earth. The armament consisted of 14 24- pounder guns.

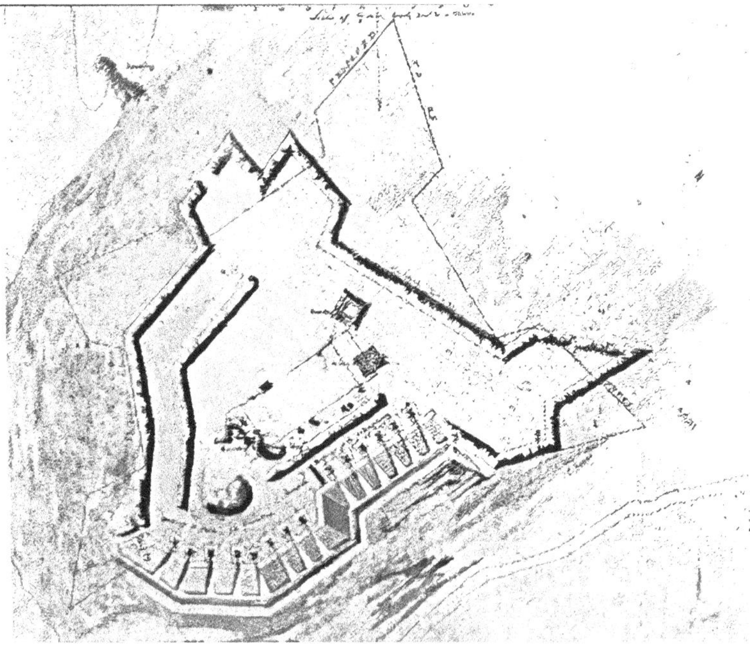

Rivardi's original plan for Fort Nelson with Latrobe's design superimposed on top of it NARA

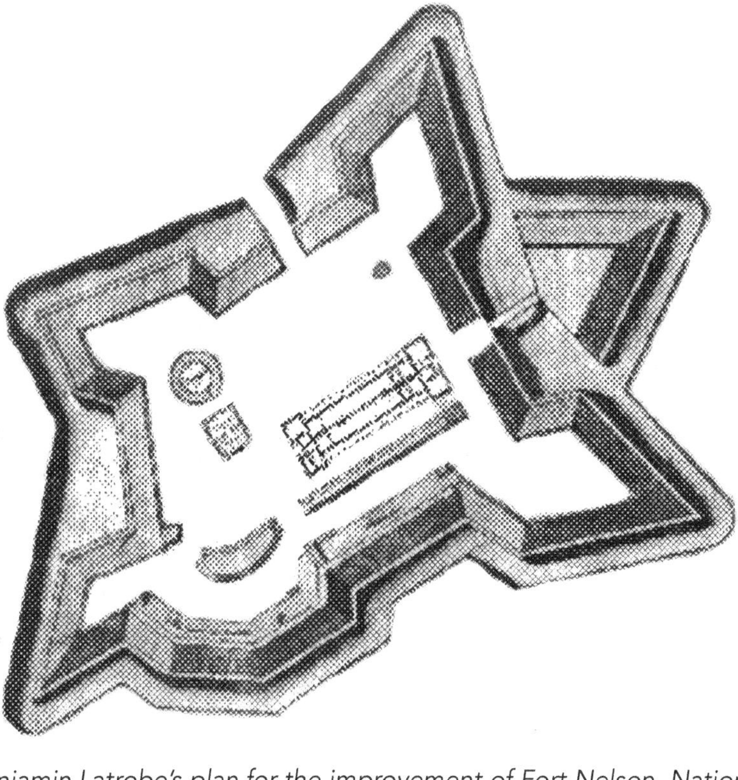

Benjamin Latrobe's plan for the improvement of Fort Nelson. National Archives

In 1798, the "Quasi War" with the French navy was beginning to intensify. That year, the great engineer/architect Benjamin Henry Latrobe (1764-1820) surveyed the fortifications of Norfolk and submitted a design for the improvement of Fort Nelson. Improvements included the addition of bastions, a wider ditch, a ravelin, and completion of traverses to shield the barracks and magazine. Latrobe's design eliminated the embrasures that could funnel cannonballs inward killing defenders. Instead, a continuous rampart was designed where guns could be mounted on carriages en barbette. The improvements were made between 1802 and 1804.

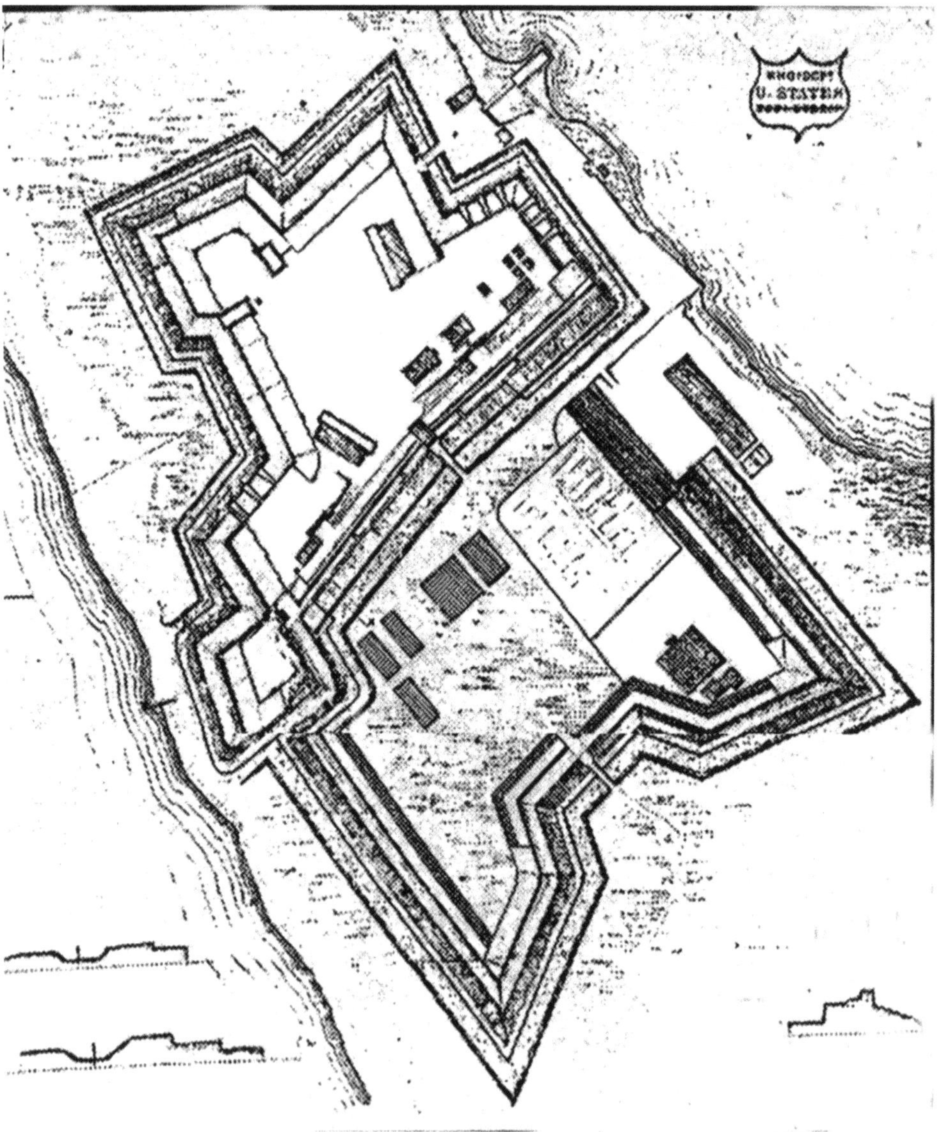

Plan of Fort Nelson drawn in the survey of 1819, showing the masonry fort of the Second System. NARA

After the clash of the United States frigate *Chesapeake* and the British frigate *Leopard* off Hampton Roads in 1807, President Jefferson initiated the "Second System" (1807-1814) of coastal defense. During 1808 both Fort Nelson and Fort Norfolk were reconstructed and updated with masonry scarps. Fort Nelson's reconstruction was completed in less than a year.

During the War of 1812, both river forts, as well as batteries and blockhouses on Craney Island, were manned and readied for war. Though a British assault was repulsed at Craney Island and never reached Norfolk, the town of Hampton was sacked. The lessons gained from this war led to a more comprehensive system of coastal defense called the "Third System" (1817-1867).

In 1818, as the Third System forts Monroe and Calhoun (Fort Wool) began to rise at the entrance to Hampton Roads, the Elizabeth River forts were delegated to a secondary status. In 1818, Fort Nelson was garrisoned by 88 men. Fort Nelson mounted 37guns. In 1823, Company G, Third Artillery, was moved to Fort Monroe. In 1824 both Fort Nelson and Fort Norfolk were put on a caretaker status. In 1827, Fort Nelson was leveled to make way for the Portsmouth Naval Hospital. The old guardian of the river, meant to be the principal stronghold, was gone, but Fort Norfolk survived.

Fort Norfolk, Elizabeth River Begun 1794

The curved channel face of Fort Norfolk Dale Manuel Collection

In 1794, Fort Norfolk was begun out of a local project. It would be added to the First System of coastal defense when 4.31 acres were purchased by the Federal government. In 1802, the War Department ordered the fort to be disarmed. In 1808, the fort returned to Federal service. In that same year the earthwork fort became an irregular enclosed work of masonry. The main feature was a curved battery defended by irregular bastions on flanks and in the rear.

The designed armament was 30 guns, but in 1809 there were only 10 guns mounted. In 1810, a two-story brick officers' quarters was constructed in the center of the gorge wall. The fort also had a brick magazine and barracks for two companies, including officers.

During the unsuccessful British attack on Craney Island in 1813, the fort stood ready to defend the river channel. By 1818, Fort Norfolk had a garrison of 50 men and mounted 30 guns.

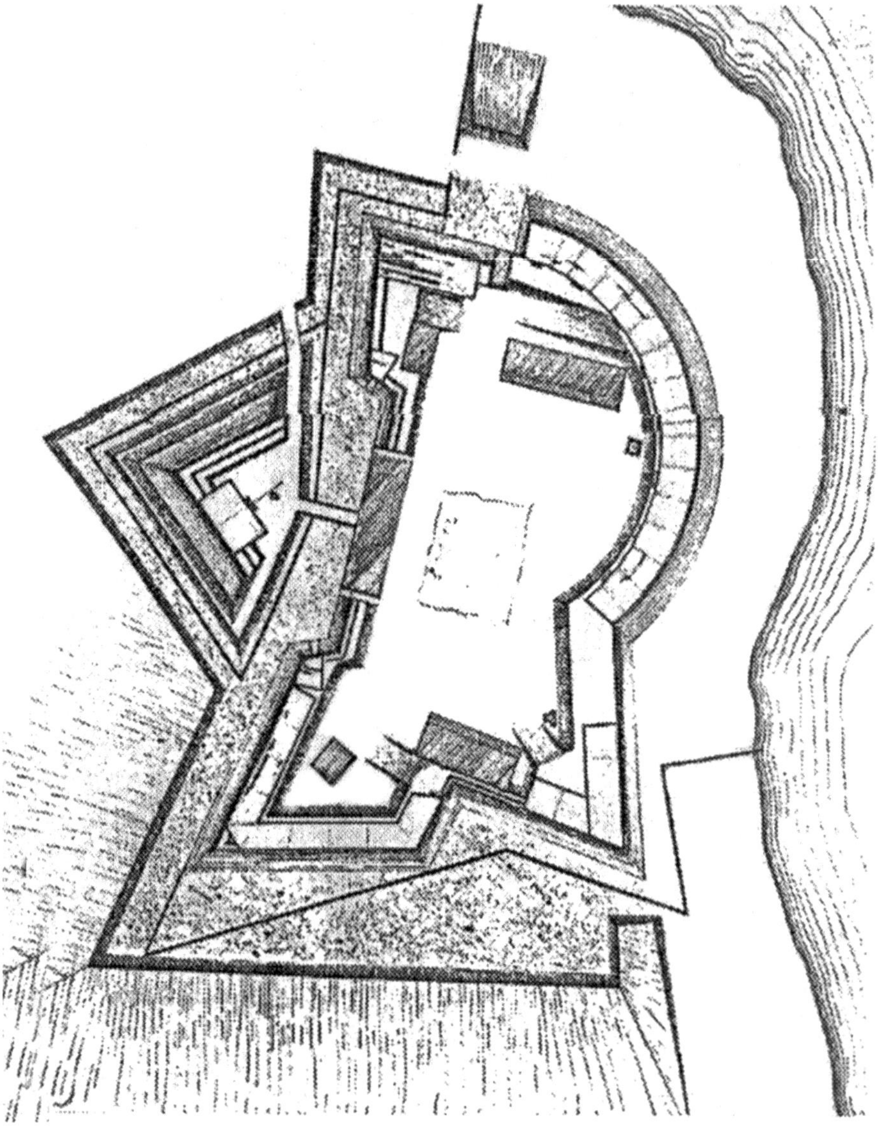

This plan of Fort Norfolk shows the curved channel face and irregular flanking bastions. The curved face offered a continuous unobstructed line of fire on the channel. In the gorge area a ravelin guarded the gorge curtain. Inside the enclosed fort were barracks and a magazine. Gun platforms lined the curved parapet and the bastion angles. Dale Manuel Collection

After 1818, the fort began to decline. It became a secondary defense with the start of construction of Fort Monroe and Fort Wool (Calhoun) near the mouth of Hampton Roads in 1819. The fort was closed in 1824. In 1849, the Army transferred ownership to the Navy with the stipulation that it not be demolished; the Navy built a large powder magazine in the center of the fort. The 1810 officers' quarter was used for ordnance operations. The property would be owned by the Navy until 1923, when

The masonry scarp and picturesque sallyport of Fort Norfolk
Dale Manuel Collection

On the parade of the fort are barracks, a storehouse and the large naval magazine. The magazine is on the left in the rear. On the right rear is the 1810 two-story officers' quarters Dale Manuel Collection

it became headquarters of the Army Corps of Engineers for the Norfolk District. Currently, the fort is one of the best surviving examples of a Second System fort with a semicircular face.

During the Civil War Confederates held the fort until Norfolk was evacuated on May 9, 1862. Union forces used the fort as a prison for the duration of the war; the 1810 officers' quarters was used to imprison captured Confederate blockade runners. Graffiti from this period remains on the walls. Restoration of the quarters has been carried out by the Norfolk Historical Society.

Photo of the earthen parapet and terreplein of the curved water battery. A wooden gun platform was once on the terreplein and a hot-shot furnace was on the other side of the earthen ramp Photo by John Weaver

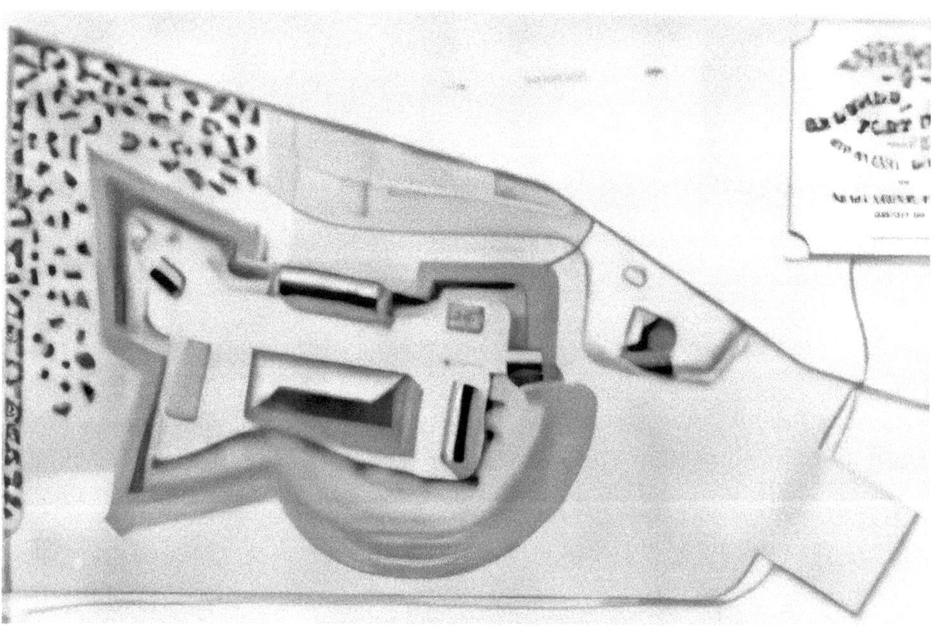

Plan of Fort Norfolk in 1860. The most prominent structure in the fort's interior is the large naval magazine. The navy gained possession of the fort in 1849 Dale Manuel Collection

The salient of the full bastion Dale Manuel Collection

A view of the gorge curtain and curtain angle. The 1810 officer's quarters made up most of the curtain wall.

Powder magazine of Fort Norfolk with its magnificent arches
Photo by John Weaver

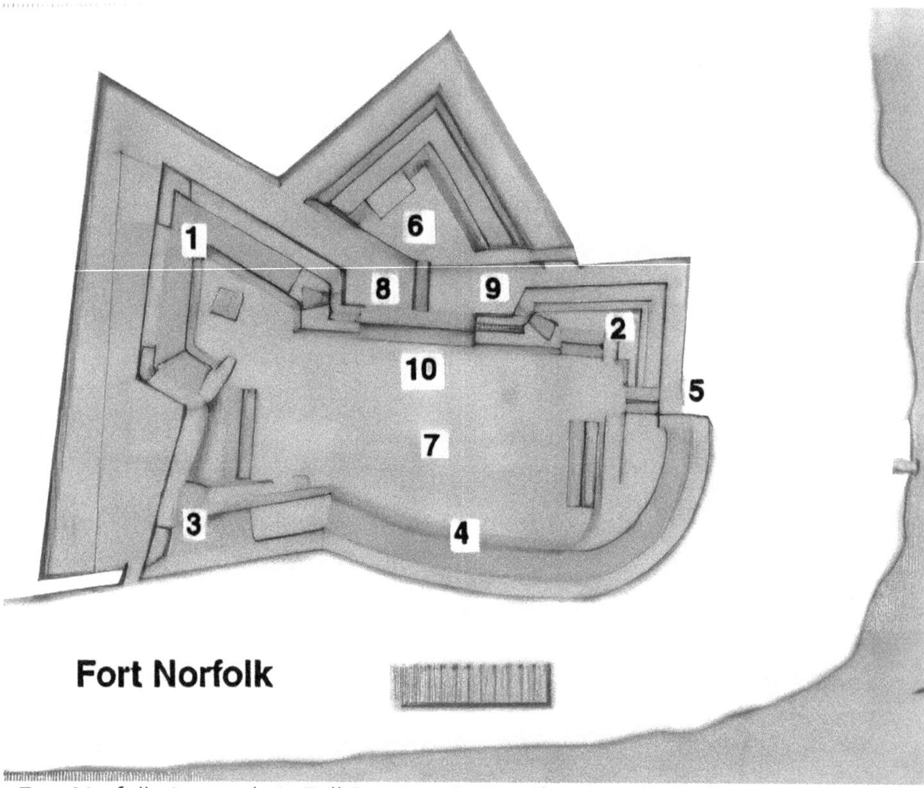

Fort Norfolk. Legend: 1. Full Bastion; 2. Demibastion; 3. Redan; 4. Semicircular Water Battery; 5. Sallyport; 6. Ravelin; 7. Parade; 8. Curtain; 9. Curtain Angle; 10. Officers' Quarters (1810) Dale Manuel Collection

Fort Powhatan, Hood's Point Begun 1808

Construction began in 1808 on Fort Powhatan on the site of the 1781 battery at Hood's Point. It was built to guard the James River approach to Richmond. Although Powhatan was the Algonquian name for the James River, the fort was probably named for Chief Wahunsonacock, whom the English called Powhatan.

Fort Powhatan consisted of a strong semicircular masonry water battery with a regular work on top of a bluff overlooking the battery. The regular work was a masonry square with sharp bastionettes in the center of three walls. The rear square work guarded the land side of the fort. There was a magazine in the square work, and one was in the water battery. Barracks and quarters were also built within the square work. A curved postern gave access from the square work to the water battery below. The fort was designed to mount 25 guns on wooden platforms in both works.

Before the War of 1812, Fort Powhatan remained about half-finished and was not armed. During the action at Hampton Roads in 1813, the Richmond defenses that included Fort Powhatan stood ready with about 4,000 men. All the fort's armament would be mounted by 1814, but the fort would see no action until the Civil War.

Sketch of Fort Powhatan Dale Manuel Collection

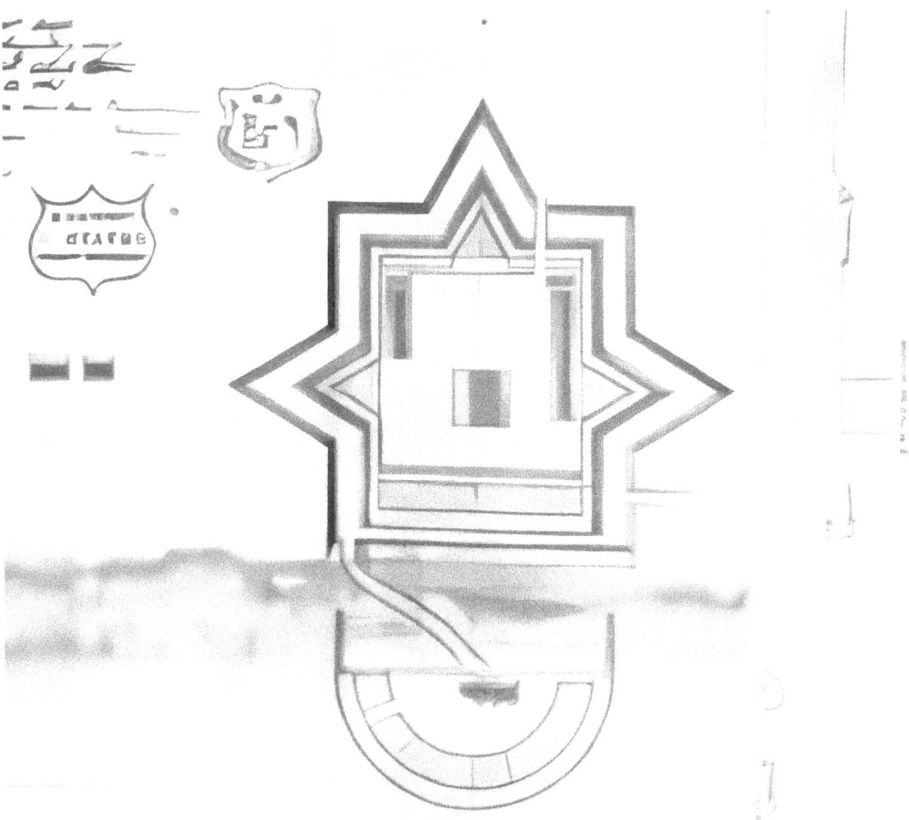

Both the lower semicircular water battery and the main works that were on top of the bluff are shown on this drawing Dale Manuel Collection

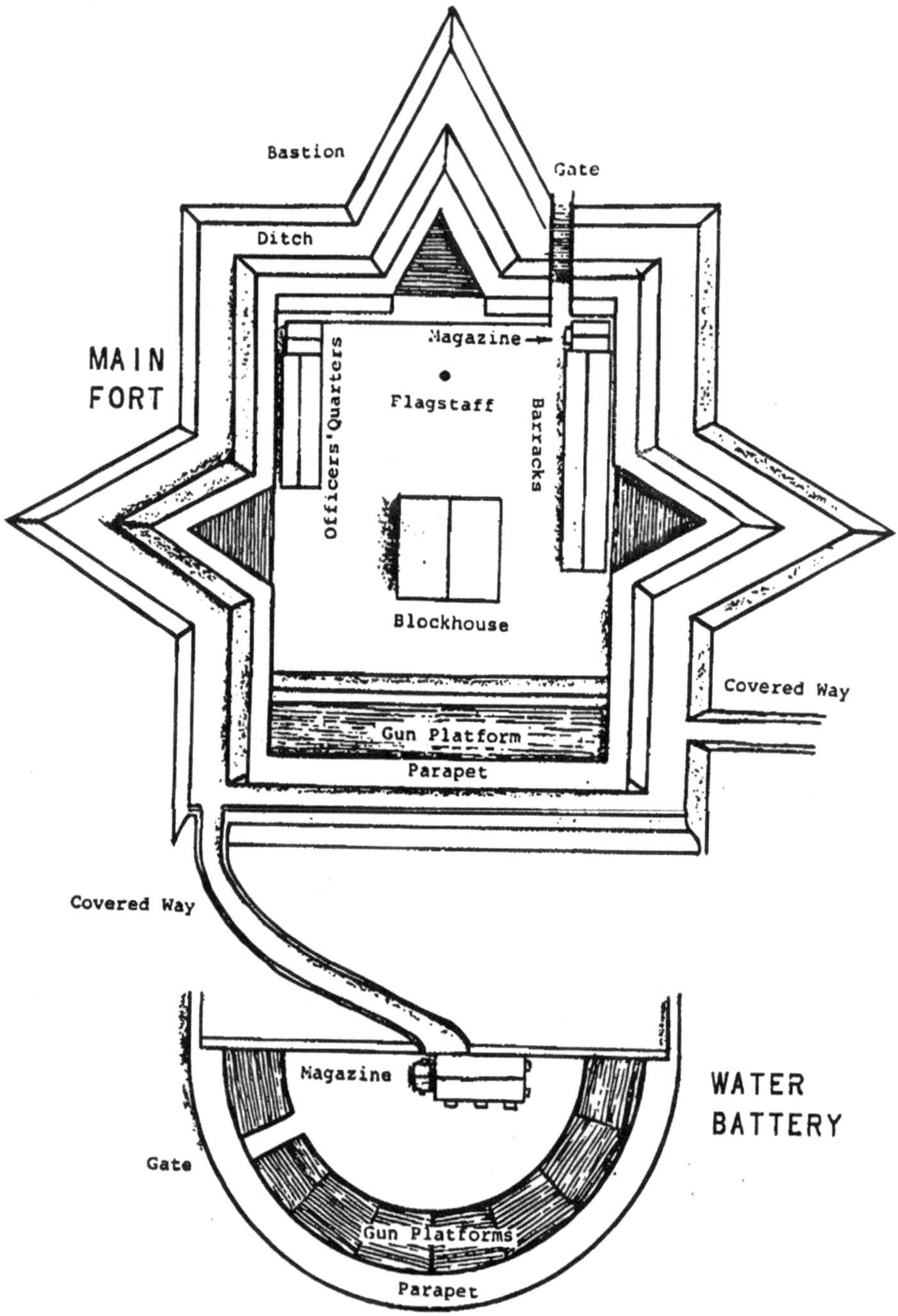

In this plan of Fort Powhatan the profile shows both magazines of the main fort and the lower water battery. Gun platforms were in both portions of the fort. Barracks and quarters were in the main fort. Dale Manuel Collection

During the coast survey of 1818, the fort mounted 13 guns. The Third System board showed no interest in the fort and it would not be regarrisoned until the Civil War.

In March 1862, Confederate President Jefferson Davis ordered that Fort Powhatan be fortified and river obstructed to prevent the ascent of ironclad vessels. In May, 1862, Union vessels under John Rodgers passed the fort in an attempt to capture Richmond. The fort's fire on the flotilla was ineffective, but the expedition was stopped at Drewry's Bluff. On the Union flotilla's return trip, Fort Powhatan's fire proved just as ineffective.

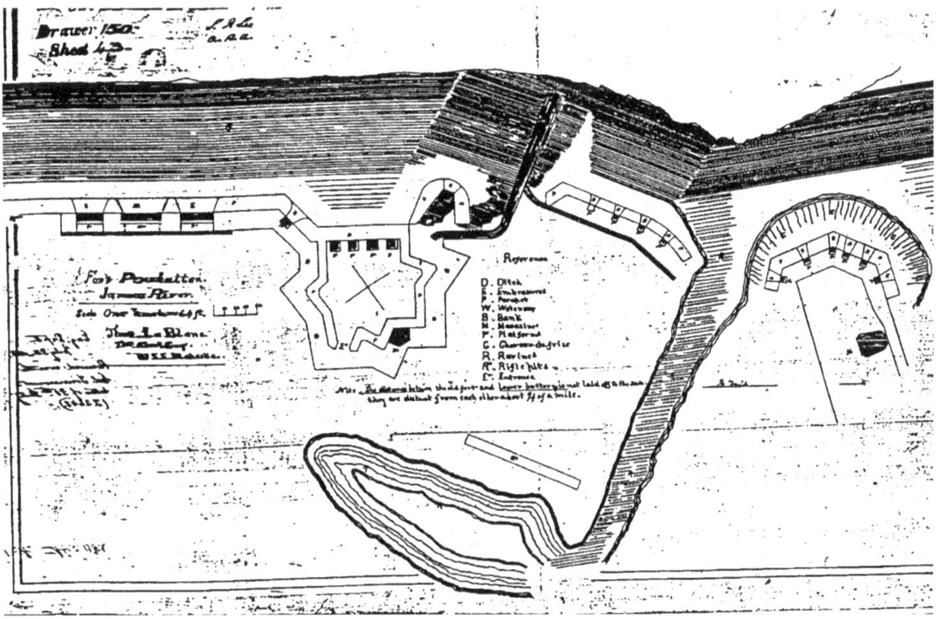

This plan of Fort Powhatan was drawn during the Civil War. The semicircular water battery is gone along with all the inner structures of the main works. A dry ditch still surrounds the main works. Nineteen gun platforms are apparent with a rifle pit in the rear. A magazine is in back of the main works with another in an adjacent battery. A chevaux-de-frise was erected in front of the adjacent battery Dale Manuel Collection

On July 14, 1863, Union Naval forces captured the fort. The forces were under the command of Rear Admiral S. P. Lee and included the vessels: USS *Sanamon, Lehigh, Mahaska, Morse, Commodore Barney, Commodore Jones, Shokoken,* and *Seymour.* Lee reported that the men had destroyed two magazines and 20 gun platforms and carriages. The last Confederate defense below Chaffin's and Drewry's Bluffs had fallen. On May 21,1864, Confederate cavalry attacked Federally held Fort Powhatan but were dispersed by gunfire from the USS *Atlanta* and the USS *Dawn.*

Some brick ruins of the fort remain on private property owned by the Prince George Hunt Club.

The Repulse at Craney Island

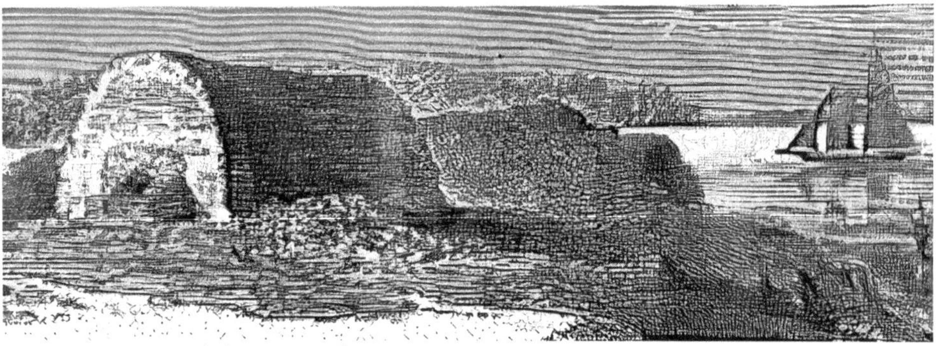

1853 view of the ruins of a magazine on the north end of Craney Island Dale Manuel Collection

During the War of 1812, both the British and the Americans recognized that Craney Island, at the mouth of the Elizabeth River, was the key to the defense of Norfolk. Norfolk was strategic because of her navy yard and because the USS *Constellation* was bottled up in her harbor. It was believed that the British would not attack Norfolk without first capturing Craney Island. The fortifications on Craney Island, the Constellation, a flotilla of 24 gunboats, Forts Tar, Barbour, Norfolk, and Nelson on opposite sides of the river were all put in the best possible state of defense in anticipation of a British attack. Brigadier General Robert B. Taylor of the Virginia Militia was in command.

Early in the war, Chief of Engineers Jonathan Williams sent the Corps of Engineers' second-highest ranking officer, Lieutenant Colonel Walker Keith Armistead, to Norfolk. His mission was to put the Norfolk defenses in the best condition possible and to develop a general defensive plan. The sending of so high an officer showed how important Norfolk's harbor was to the war effort. Armistead believed Norfolk should be defended from the advanced position of Craney Island. In June, 1813, the work

Section of the blockhouse on the north end of the island Dale Manuel Collection

Section of the blockhouse in the fort Dale Manuel Collection

Plan of the ramparts of the channel-side fort Dale Manuel Collection

on Craney Island began. The entire island of Craney was transformed into an armed camp. A battery ran the length of the island with a blockhouse on the 'north end, a redoubt in the center, and a fort on the channel side. A large blockhouse would be added later to the channel side fort. Barracks were constructed in the rear. The north battery alone contained one 18-, two 24-, and four 6-pounders to fire on the narrow strait that separated the island from the mainland. Armistead also placed another battery at Lambert's Point on the opposite side of the river mouth. The Americans only numbered 737 men to oppose 4,000 British.

As an attack became imminent, an alleged quote by British Admiral Sir George Cockburn was widely circulated among the American troops. He was said to have boasted, "We will storm Fort Nelson and be in Norfolk to supper. There you will find two banks and for your exertions and bravery you will have three days' plunder and the free use of a number of fine women." Whether the quote by the British admiral was said, the Americans wanted to believe it and did; the hot-shot furnaces were stoked around the clock.

On June 21, 1813, a small flotilla of sloops and schooners formed a protective net at the river's mouth. Among the sailors of the flotilla was a young midshipman named Josiah Tattnall, who would one day return to Norfolk to command the CSS *Virginia.* At 9:00 am, 2,500 British landed on the mainland opposite the north end of the island. Taylor's batteries hit them so hard they were forced to retreat into the woods. Simultaneously, 50 barges carrying 1,500 men approached the island.

The barges were led by the admiral's beautiful barge called the *Centipede.* It was so named because of its length and great number of oars on each side that looked like legs. In her bow was a brass 3-pounder, called a "grasshopper," and she was commanded by Captain Hanchett, a natural son of King George III. As the barges drew near to

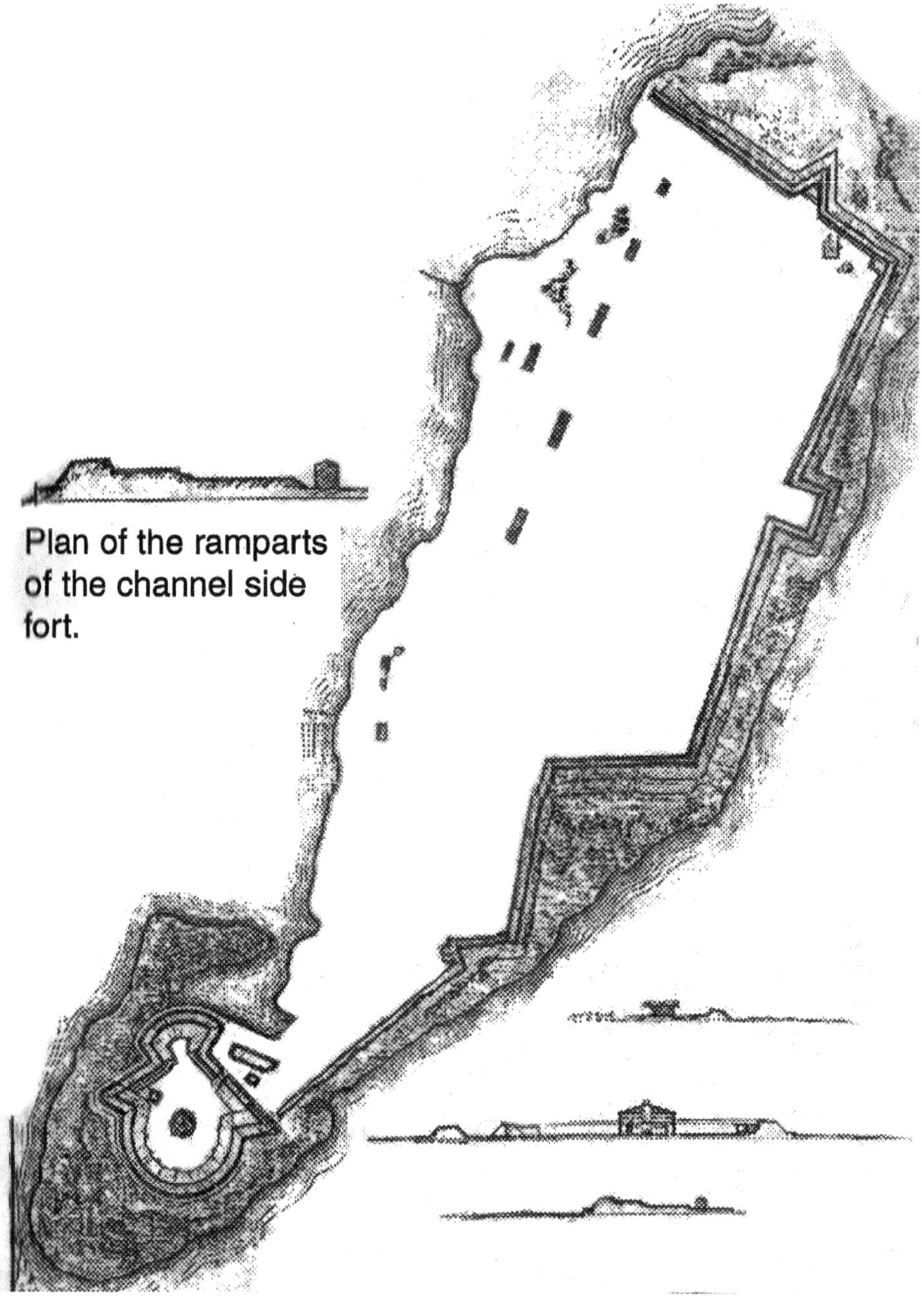

1819 survey plan of the fortifications on Craney Island Dale Manuel Collection

1853 view of the blockhouse that was in the center of the channel-side fort Dale Manuel Collection

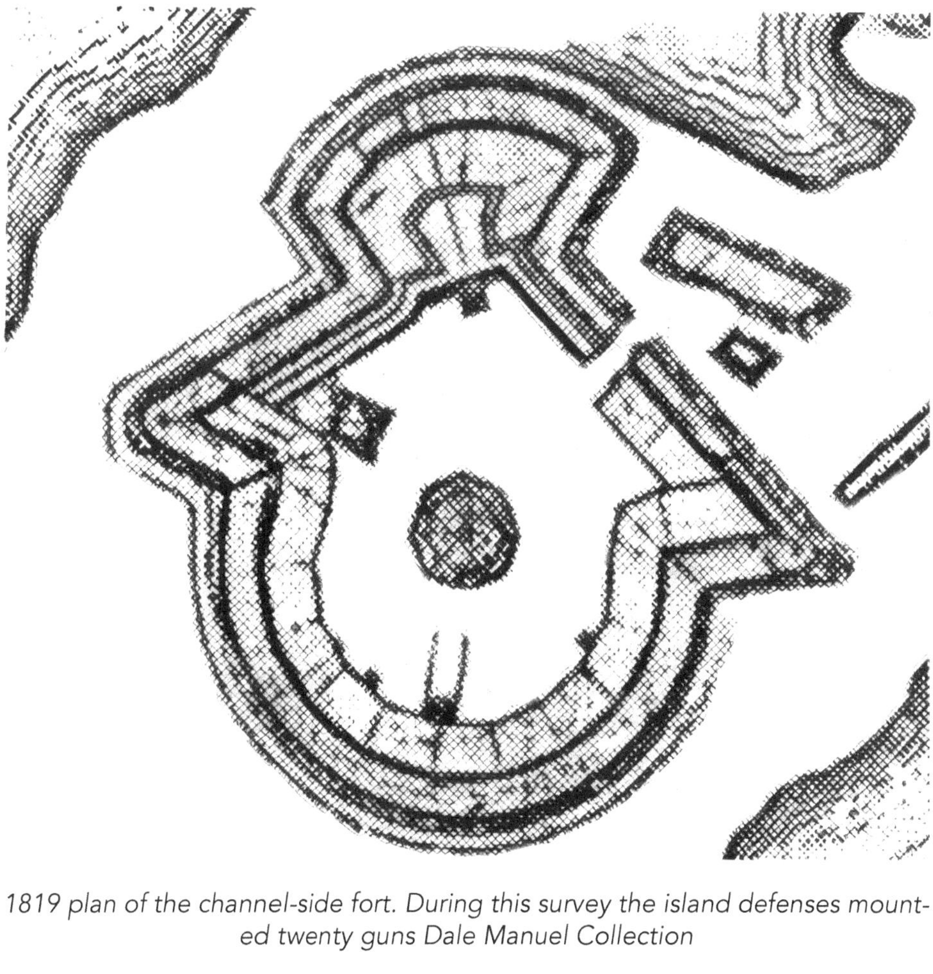

1819 plan of the channel-side fort. During this survey the island defenses mounted twenty guns Dale Manuel Collection

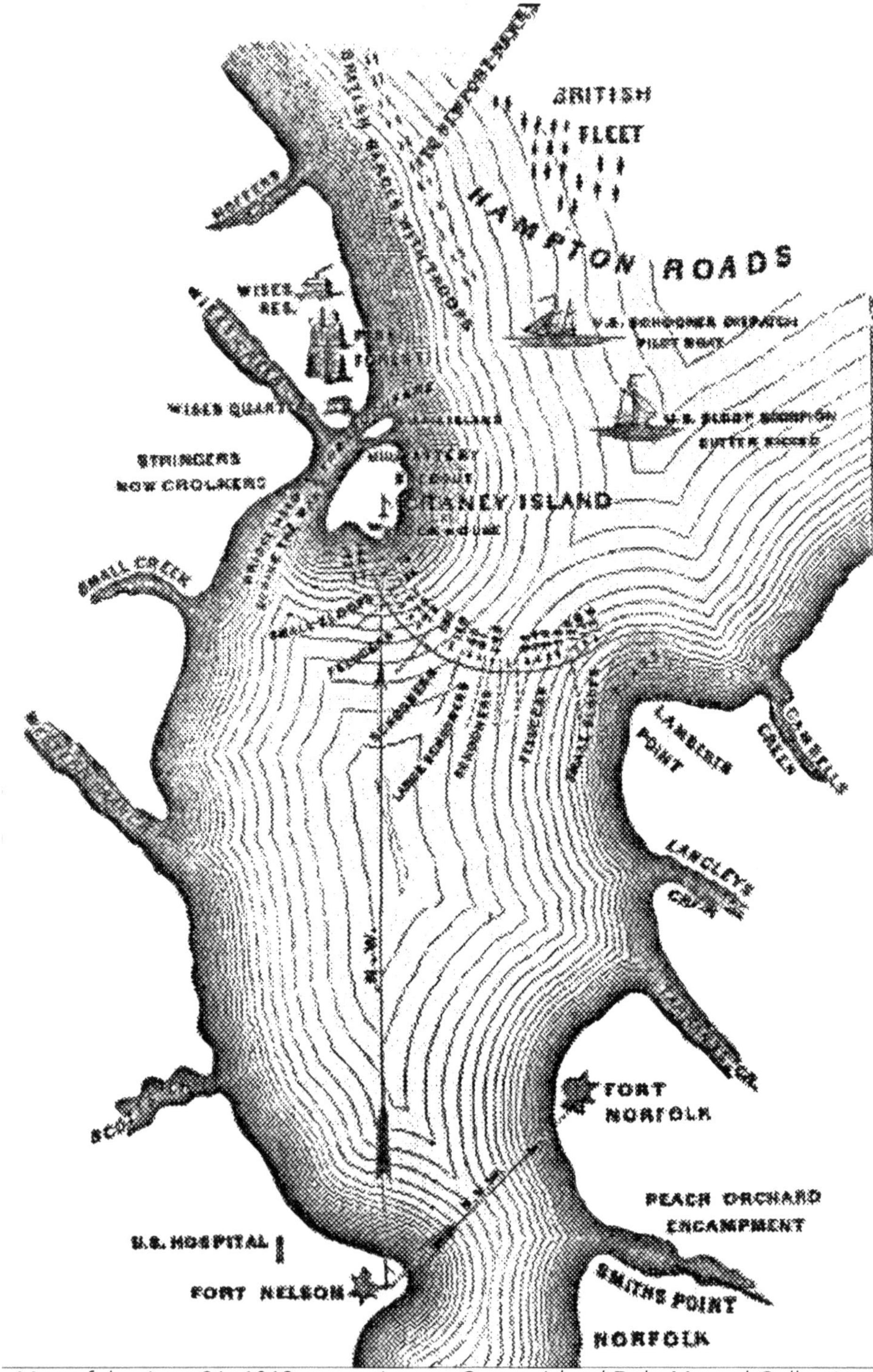

Map of the June 21, 1813, operations at Craney Island Dale Manuel Collection

British Admiral Sir George Cockburn is said to have boasted "We will storm Fort Nelson and be in Norfolk to supper. There you will find two banks and for your exertions and bravery, you will have three days' plunder and the free use of a number of fine women." Whether the quote was said or not, the Americans wanted to believe it and did. Dale Manuel Collection

the island, the whole battery, except two guns that had been dismounted, belched fire and smoke and hurled round, grape, and canister shot. The volley was fearsome with five of the barges hit and sank. The *Centipede* was hulled by a round shot that passed through her, wounding several of her men, cutting off the legs of one of them, and severely wounding Captain Hanchett.

The other boat crews were thrown into great confusion as the water filled with Englishmen swimming for their lives. Some of the Americans waded out into the water to capture prisoners. The remainder of the barges escaped back to the fleet.

Tattnall and others salvaged the half-sunken *Centipede* and after it was repaired it was used by the Americans as a guard-boat. The Americans did not lose a single man. The British had six killed, 24 wounded, and 114 missing. Although Admiral Cockburn would continue waging his amphibious warfare on the Southern Coast, Norfolk would be safe for the duration of the war.

The Repulse at Craney Island was important because it demonstrated what one capable engineer could do with a well-planned defense, and that forward positions would better prevent attack. Despite the embarrassment of the burning of Washington during the war, the defenses of Craney Island; Fort McHenry, Maryland; Fort St. Philip, Louisiana; and Fort Bowyer, Alabama, all proved coastal forts could provide successful defense. They all helped pave the way for the establishment of the Third System.

The Third System board showed little interest in Craney Island. Although it was listed to be preserved it was abandoned until the Civil War. During the Civil War the Confederates constructed a large 42-gun battery on the island. The island served as the anchorage for the CSS *Virginia.* Off its shore, the ironclad was burned. The battery was evacuated on May 9, 1862, along with the other defenses of Norfolk. Little remains of the defenses, the enlarged island is now a peninsula and is still a military reservation.

The Confederate garrison on Craney Island cheered as the CSS Virginia steamed out to meet the Union fleet Dale Manuel Collection

A drawing of the forty-two gun Confederate battery on Craney Island.

Defenses of Savannah, Georgia

Legend: 1. Fort Jackson; 2. Tybee Island; 3. Wassaw Island; 4. Wassaw Sound; 5. Ossabaw Sound; 6. Fort Pulaski; 7. Bird Island; 8. Venus Point; 9. Fort McAllister; 10. Tybee Roads; 11. Caustern's Bluff; 12. King's Bridge; 13. Vernon Rive;r 14. Wilmington River; 15. Fort Thunderbolt; 16. Whitemarsh Island; 17. Skiddaway Island; 18. Battery Daniels; 19. Fort Wimberly; 20. Atlantic & Gulf Railroad; 21. Bonnaventure Battery; 22. Great Ogeechee River

Fort Wayne, Savannah Rebuilt 1812

In 1808, Army engineer Alexander Macomb submitted a plan that would replace an older earthwork fort that had been known as Fort Wayne. Macomb's plan called for a fort with a semicircular water battery, bastions, and even ravelins. His plan was

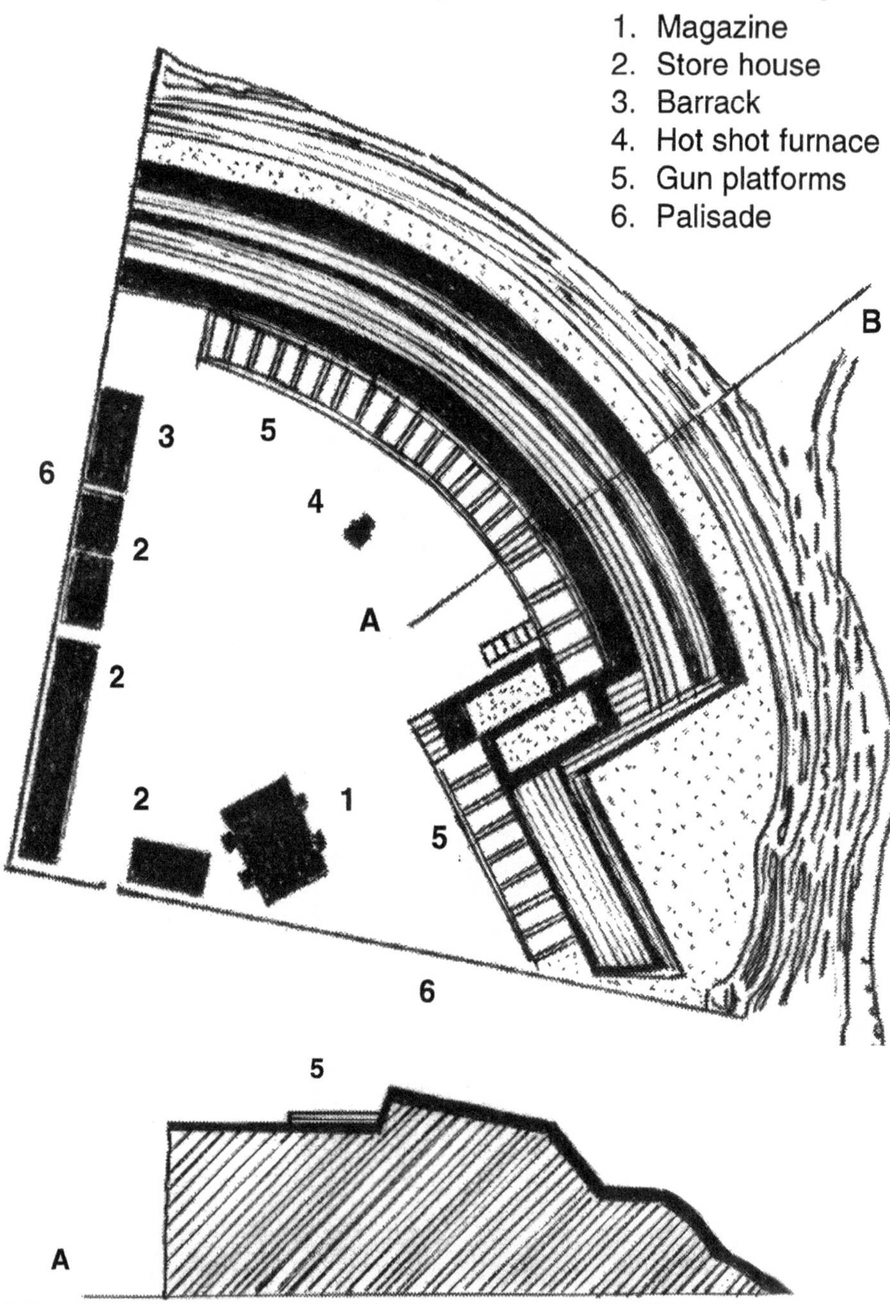

a) Plan b) section of a portion of the Second System Fort Wayne from the survey of 1821, William McRee's design is shown. Dale Manuel Collection

never adopted. Instead, in 1812, William McRee began the Second System fort that was to have a semicircular water battery, but only a palisade would enclose the rear of the work. The fort was rebuilt on the site of two older British forts, Forts Halifax and Prevost, in an area of town known as Trustee's Garden. Situated on a bluff that overlooked the river, its water battery was strategically located.

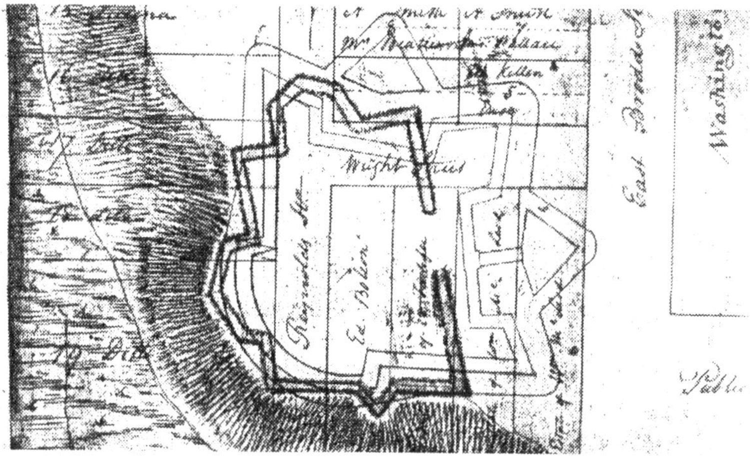

Macomb's 1808 plan for Fort Wayne superimposed on the old earth fort.
Dale Manuel Collection

Behind the earthen rampart and brick revetment were wooden gun platforms. The palisade ran from the battery and enclosed the parade. Barracks, storehouses, and a magazine lined the inside of the palisade. Behind the battery was a hot-shot furnace.

The fort was named in honor of Major General "Mad" Anthony Wayne. He served at the battles of Brandywine, Germantown, Monmouth, Stony Point, Jamestown Ford, and Yorktown during the Revolutionary War. From 1791-1792, he served a term in the U.S. House of Representatives from Georgia. On August 20, 1794, as a major general and commander-in-chief of the Northwest, he was victorious at the Battle of Fallen Timbers in northern Ohio.

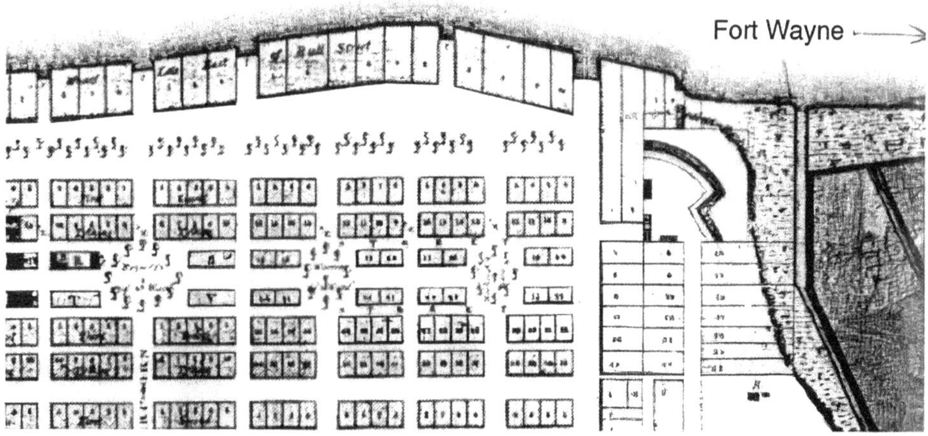

In this map of Savannah, dated 1820, the site of Fort Wayne is shown
Dale Manuel Collection

Drawing of Fort Wayne Dale Manuel Collection

During the War of 1812, Fort Wayne was served on a rotating basis by the same companies that served Fort Jackson. Fort Wayne was the terminus of a signal or early warning system that was initiated from the lighthouse on Tybee Island, then relayed to an observation post on Cockspur Island, and then relayed by Fort Jackson to Fort Wayne in the city.

When constructed, Fort Wayne was a secondary defense, while further downriver, Fort Jackson was a primary defense of the Second System. As the Third System fort that would become known as Fort Pulaski began to rise out of the mud of Cockspur Island in the late 1820s, Fort Jackson became a secondary defense, and Fort Wayne became obsolete. The trend was to build forts further away from the cities they defended and closer to the sea. Eventually, the fort would be razed, and the Municipal Gas Plant would be built on its site.

Fort Jackson, Savannah River Begun 1808

In 1808, construction was begun on Fort Jackson, three miles east of Savannah on the south side of the Savannah River. Built on the site of an earlier Revolutionary War earthen battery, this strategic location guarded a stretch of the river where all channels converged. It was also near a deep-water anchorage known as five-fathom hole.

The fort was named in honor of Revolutionary War leader James Jackson. Jackson led the Georgia Militia in the pivotal battle of The Cowpens in South Carolina on January 17, 1781, where the British, under Lieutenant Colonel Banastre Tarleton, were routed. As a Lieutenant Colonel, Jackson accepted the keys to the city of Savannah from departing British troops on July 11, 1782. Jackson would later be elected Governor of Georgia and a United States Senator.

Because the fort was built on a marsh, its brick scarp foundation was laid on top of wood pilings and a grillage of timbers and planks. As long as the wood remained submerged, it would not rot. The ramparts of the water battery were constructed with 140,000 bricks. The supervising Army engineer, William McRee, had earth rammed

The water battery from the parade of Fort Jackson Photo by John Weaver

The water battery, one of the flanking structures, and the sallyport of Fort Jackson from across the ditch Photo by John Weaver

Katina Lear

Confederates once again man and fire the thunderous 32 pounders
Dale Manuel Collection

One of the flanking structures that controlled a landward curtain
Photo by John Weaver

The main magazine was constructed in the 1850s. It is entered from inside the gallery. The inside is vaulted with the capacity to hold 10,000 pounds of powder. On the left, out of the picture, is the sallyport Dale Manuel Collection

between the walls of the battery. A wooden gun platform and brick magazines were built behind the ramparts. Also behind the battery was a hot-shot furnace. A wooden palisade would enclose a wooden barracks in the rear of the work.

The fort was completed by the start of the War of 1812. During the war, three companies provided defense for Savannah and Fort Jackson: the Chatham Artillery,

The outside of the water battery and wet ditch. This is the oldest part of the fort Photo by John Weaver

The gorge of Fort Jackson showing the bridge spanning the wet ditch and the two bastions protecting the sallyport Photo by John Weaver

the Republican Blues, and the Savannah Heavy Artillery. The 8th U.S. Infantry also supplied companies to the Savannah defenses.

The current fort has an odd shape that is a combination of an older Second System water battery and palisade that was replaced with a surrounding Third System masonry wall. Flanking fire was provided by the addition of four demibastions in the masonry walls.

Improvements to the fort were made between 1845 and 1861. Construction lagged on for years due to the fort being evacuated during the warm months of the year because of the lethal fevers the surrounding marsh spawned. Behind the semi-

Civil War drawing of Fort Jackson and Battery Lee Dale Manuel Collection

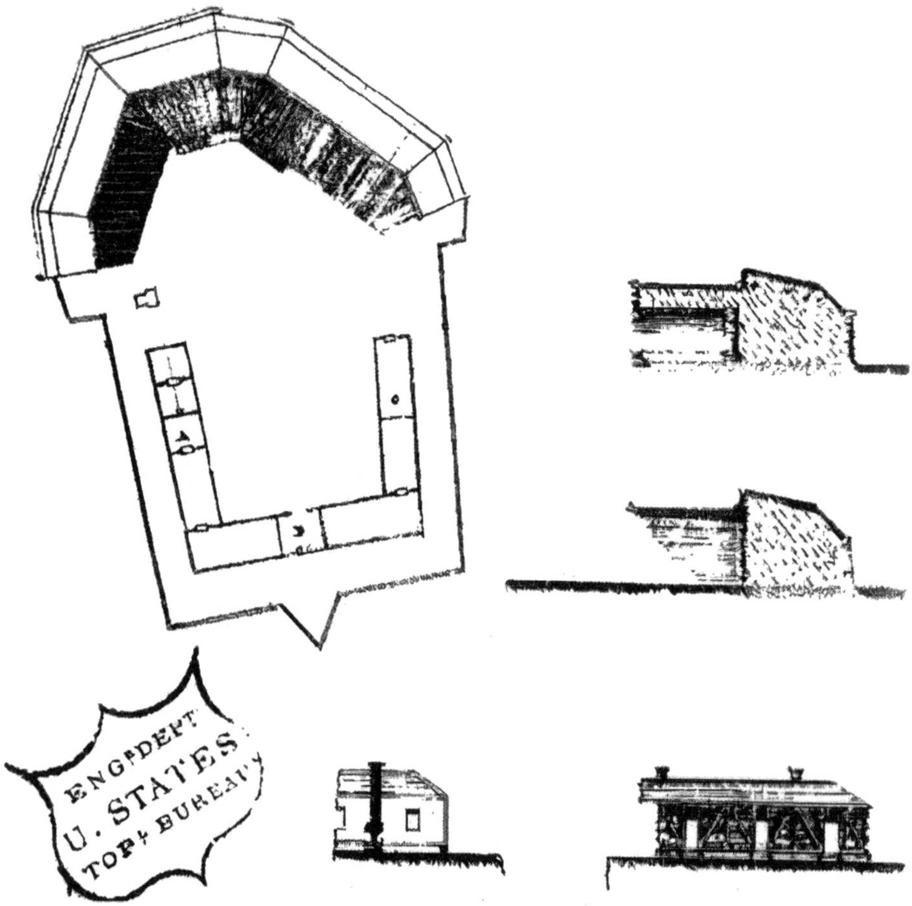

Plan and sections of the Second System works at Fort Jackson
Dale Manuel Collection

circular rampart of the water battery, a two-tiered terreplein was built to mount guns en barbette. Under the terreplein a vaulted brick gallery supported the weight of the guns and was used as storage and support space. Although the flank guns were never mounted in the demibastions, the scarp and ditch could be swept of attacking infantrymen by firing through the embrasures with 24-pounder howitzers on special carriages. A sallyport was constructed just past the end of the battery, opening with a drawbridge across the ditch.

Musket loopholes line both flanks. At the end of the flanks are demibastions that flanked the gorge curtain. Inside the fort, barracks were constructed on both sides of the parade, inside the masonry curtains. One barracks was designed to house one company of enlisted men and contained sleeping quarters, kitchens, and mess areas.

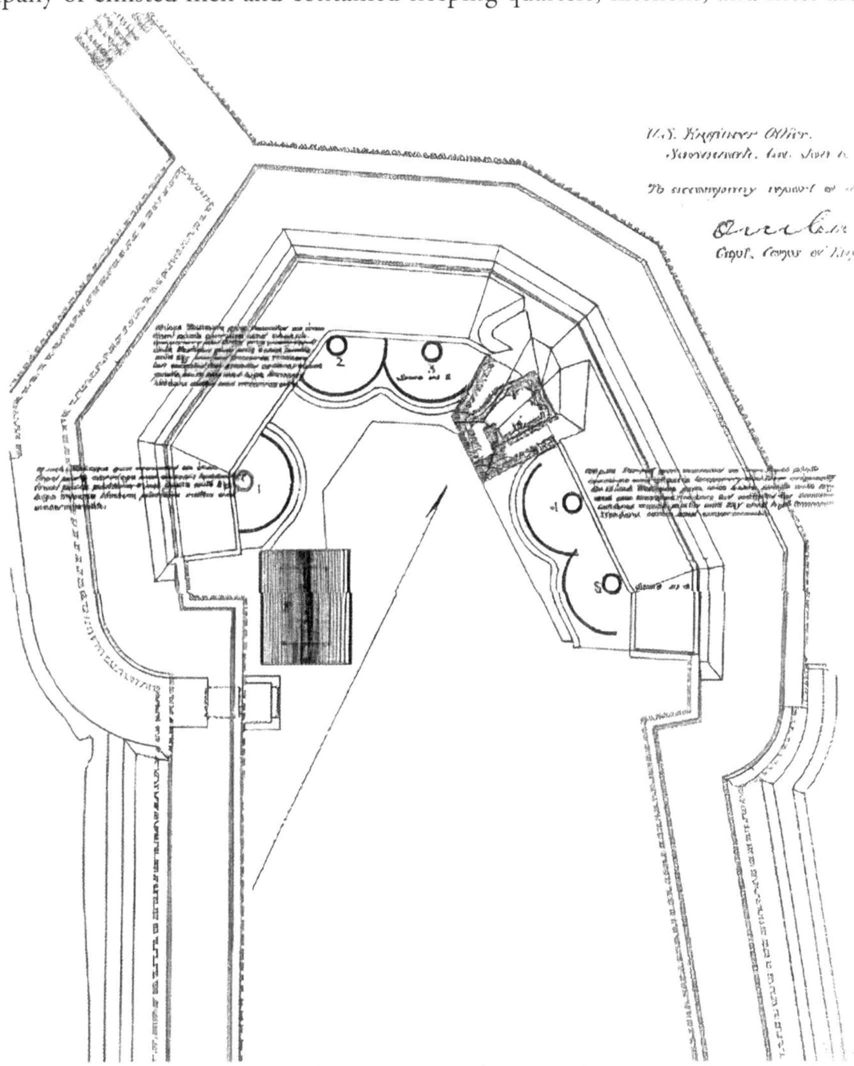

The 1896 survey of the water battery area of Fort Jackson. On the terreplein is the traverse magazine and emplacements for three 10-inch Rodman guns and two 100-pounder Parrot rifles. Dale Manuel Collection

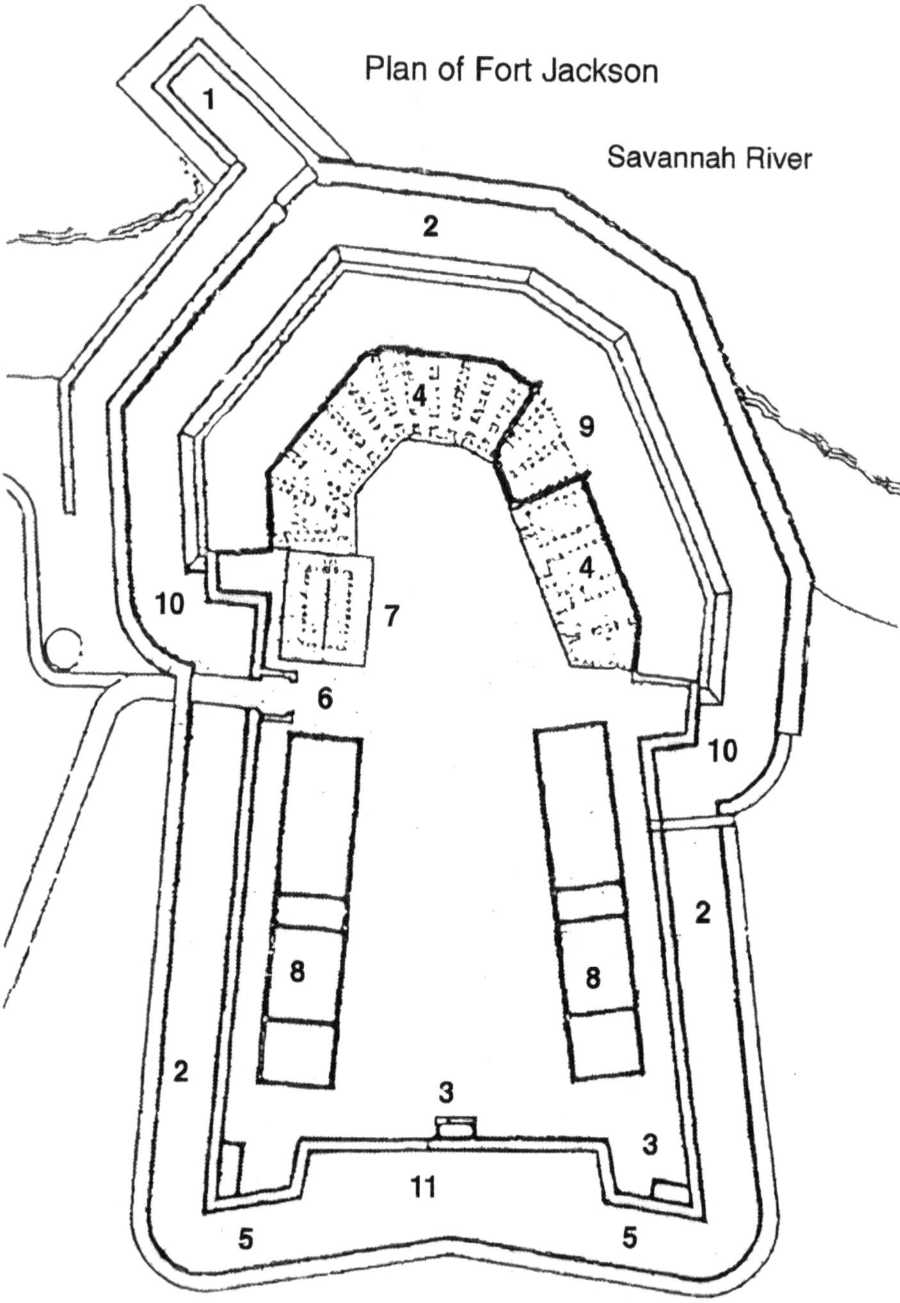

Plan of Fort Jackson
Fort Jackson plan. Legend: 1. Tidal tunnel; 2. Moat; 3. Privies; 4. Terreplein; 5. Bastions; 6. Sallyport; 7. Main Magazine; 8. Barracks foundation; 9. Traverse magazine; 10. Flanking Structure; 11. Gorge
Dale Manuel Collection

An officers barracks was under construction when the Confederates seized the fort. As the Confederate garrison grew in size, a kitchen was erected between the barracks to prepare meals.

Privies were built along the gorge wall. A main magazine was built between the sallyport and the battery. The slate-roofed, brick-vaulted magazine could store up to 10,000 pounds of black powder. The entire works are surrounded by a wet ditch that is filled by the river through a tidal tunnel. A gate retained the water in the ditch at low tide. A mechanism raised and lowered the drawbridge; when raised, the drawbridge would bar entry into the sallyport. On the inner side of the sallyport, large doors barred entry into the parade. Any enemy infantry caught inside the sallyport could be fired on from each side. Smaller posterns are in the opposite flank scarp and in the center of the gorge curtain.

Compared to the massive walls of Fort Pulaski, only about ten miles downstream, the flank and gorge walls of Fort Jackson seem weak. This is perhaps because there was not that much concern for land side defense when the area around the fort was covered with marsh and rice paddies.

Besides William McRee, who would go on to be a major and a member of the Third System Board of Fortifications, many other distinguished engineers served at Fort Jackson. Captain Joseph Mansfield was in charge of Fort Jackson as well as Fort Pulaski. He surveyed Fort Jackson in 1845, before leaving for Texas. In 1845, Lt. B.S. Alexander began new work on the fort. Starting in 1849, Captain Jeremy Gilmer su-

The sallyport of Fort Jackson
Photo by John Weaver

The gallery under the terreplein
Photo by John Weaver

pervised construction for ten years. In 1858, Lt. William H. C. Whiting was placed in charge of construction. During the Civil War, many of these engineers would become generals in the armies, both North and South. Some would die during the war.

On January 26, 1861, Fort Jackson was seized by Georgia State troops under the orders of Georgia Governor Joseph E. Brown, to prevent it from being occupied by Federal troops. The Civil War was about to begin. During the early months of the war the fort's garrison was made up of local militia units, such as the Republican Blues. On November 26, 1862, the 22nd Battalion of the Georgia Artillery was formed to garrison the Savannah River Batteries. In command of Fort Jackson and the Savannah River battalions was Colonel Edward C. Anderson. Fort Jackson was headquarters for the Savannah River defenses during the war. With the fall of Fort Pulaski in April 1862, Fort Jackson and nearby batteries became the first line of defense of the main water approach to the city.

Parade of Fort Jackson Photo by John Weaver

Exterior of sallyport of Fort Jackson Photo by John Weaver

During the war, shell-blocking earthen traverses were erected on the terreplein to protect the gun batteries. The Confederates added an armor shield of railroad iron to the east face of the main magazine. The old War of 1812 magazines were used as service magazines and shielded as well. Eight guns were mounted on the parapet of the fort. They consisted of five 32- pounder smoothbores, two 8-inch columbiads, and one 32-pounder rifle. A spur of the Atlantic and Gulf Railroad was laid to Fort Jackson and a nearby earthen battery, Battery Lee. The spur was used to bring heavy guns, supplies, and building materials.

In the first three and one-half years of the war, the Savannah River defenses kept a vigilant watch toward the sea. In December 1864, a new threat approached Savannah, not from the sea, but from the land, as General William T. Sherman's Army of 60,000 approached Savannah on their destructive "March to the Sea." By December 10, 1864, the city was besieged. The Confederates had only about 10,000 troops to oppose the siege. Confederate General William J. Hardee ordered an evacuation of Savannah and its river defenses on the night of December 20, 1864. The garrisons of

Distant view of the gorge of Fort Jackson Photo by John Weaver

Gorge curtain of Fort Jackson, showing loopholes for musketry
Photo by John Weaver

the river batteries assembled at Fort Jackson prior to leaving. The men threw arms into the wet ditch, spiked the guns, set fire to the barracks, and booby-trapped the magazine. They were transported by steamer to Screven's Ferry and marched to Hardeeville, South Carolina, the same night.

When Savannah was occupied by Federal troops, two regiments were dispatched to occupy Fort Jackson. The 28th Pennsylvania Veteran Volunteers and the 29th Ohio Volunteer Infantry under the command of Colonel John Flynn arrived at the fort at 7:00 A.M. on December 21, 1864. The barracks were still burning. After Flynn's men raised the flag over the fort, the ironclad ram CSS *Savannah* started firing at the fort. When Union batteries in Savannah began firing at the ram, a gun duel ensued.

Civil War view of Battery Lee and Fort Jackson from the Savannah River
Dale Manuel Collection

During the early 1870s, a plan was made to upgrade and strengthen the position by the addition of 15-inch Rodman guns. Instead, three 10-inch Rodman guns and two 100-pounder Parrot rifles were added to the water battery of the fort. A traverse magazine was built on the terreplein. An 1896 survey of the fort, then named Fort Oglethorpe, mentions that the timbers of the traverse platforms were rotten and unusable.

The fort was put on a caretaker status until 1905 when the property was sold. The fort lay abandoned and at the mercy of the elements until the 1960s. The fort was eventually restored and administered with excellent stewardship by the Coastal Heritage Society. The fort operats as a museum and on special occasions features uniformed reenactment groups that fire the thunderous 32-pounder guns.

Martello Tower, Tybee Island Begun around 1815

This round tower was built close to the 1773 lighthouse on Tybee Island to guard the mouth of the Savannah River. Built around 1815, the tower was constructed of tabby, a mixture of sand, shells, and cement made from shells. It was known as the Martello Tower after the famous tower that was located on Cape Martella in Corsica. In 1794, the British had great difficulty capturing this tower from the French garrison. The design of these towers provided 360 degrees of fire.

Martello Tower Tybee Island. John Weaver Collection

The Tybee tower was designed by Savannah architect Isaiah Davenport. It was four stories tall with a roof on top. The circular walls were 30 feet high and 10 feet, 8 inches thick. The tower's diameter at its greatest point was 48 feet. The basement story contained a magazine. One plan shows a 50-foot well, and another plan shows a cistern in the bottom floor. A center column made of brick and timber rose from the basement foundation and supported the upper floors. The flooring was made of wooden planks supported by timbers. The basement had a door to the outside.

A staircase led to the second story that had a door to the outside and a staircase to the third story. The second story was used for storage, while 12 loopholes circled the third-story wall. The third story had three fireplaces, and a staircase led to the top

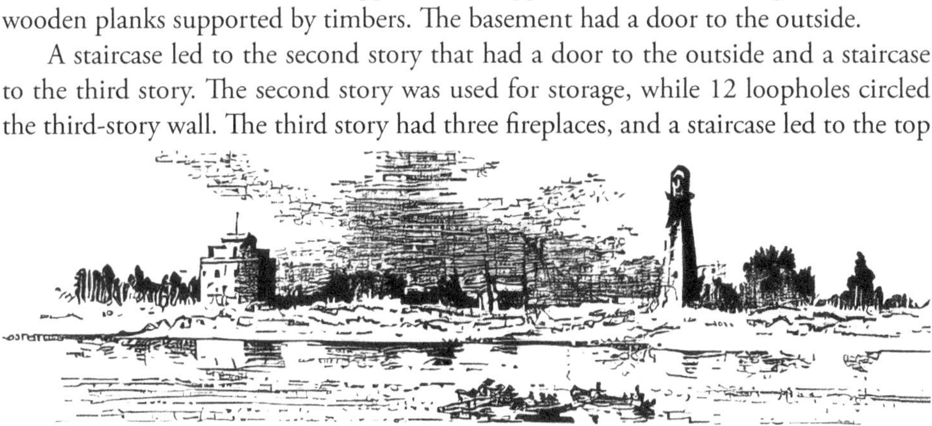

Civil War drawing of the Tybee Island Lighthouse and Martello Tower
Dale Manuel Collection

platform. The top platform once held a single gun on a 360-degree pivot. Crenels were cut into the circular tabby parapet through which the gun could fire. A wooden machicoulis encircled the top platform and protruded from it. From the loopholes in the machicoulis, muskets could be fired on attackers below.

Photograph of the tower during the Civil War. The roof is missing and a wooden structure was added to the top of the tower John Weaver Collection

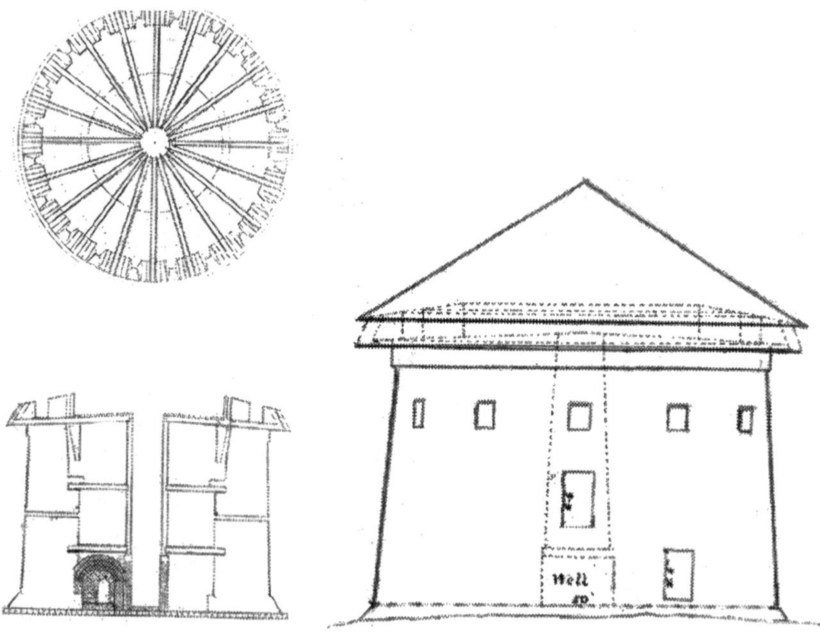

Plan, section, and elevation of the tower, including the original roof Dale Manuel Collection

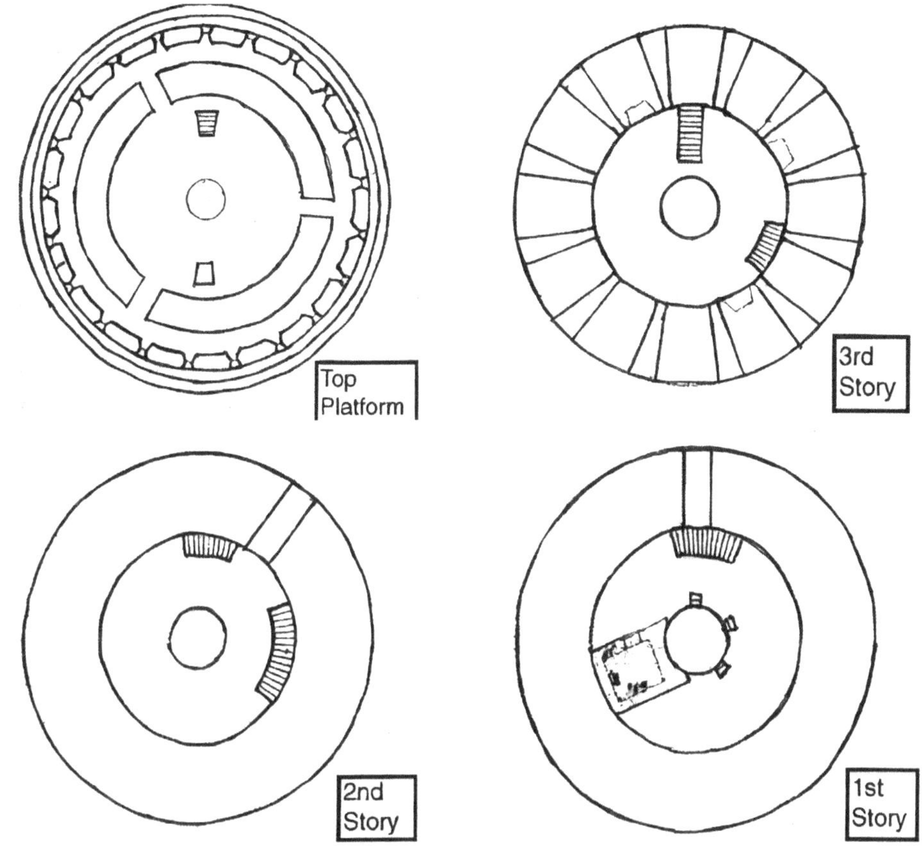

Plans of various levels of the Tybee Island Martello Tower
Dale Manuel Collection

Lieutenant Joseph Mansfield surveyed the tower in 1835. His report indicated that the tower needed major repairs. Although the magazine was dry, the wood lining was rotten. Both outside doors were nonfunctional. The floors of the second and third stories were sound and good. The platform of the wooden machicoulis was rotten and needed renewing. The roof that had been put on within ten years of his report had sound rafters but needed new shingles. Lt. Mansfield recommended that a lock be put on the outside door of the first story. He also mentions that some old cannon could be seen lying about the tower, as well as the remains of the still-standing chimneys of the long-gone company barracks. Lt. Mansfield made one comment that was prophetic when he said, "The nature of tabby is such that the tower is gradually wasting away. Fortunately, in this case, the walls are thick and sound; it will stand for years and answer the purpose intended." The tower lasted into the twentieth century when the Army razed it. Joseph Mansfield, known as the builder of Fort Pulaski, would become a Union General during the Civil War and die at the Battle of Antietam.

The Coastal Fortifications Board that planned and executed the Third System used towers as a secondary defense or as a defense of a less important location than those that were protected by major forts. Also, the development of steam-powered warships rendered a single cannon in a tower insufficient. To remedy this, plans were developed to add a battery to supplement the armament of the tower, but the battery was not constructed and the site was not garrisoned until the Civil War.

The site around the tower became a dueling ground during the antebellum period (before the Civil War). Since dueling was outlawed in South Carolina, disputes were settled on this field of honor on the Georgia side of the river. Many Carolinians lost their lives here.

In the summer of 1861, Confederates erected batteries on Tybee Island. The lighthouse was used as a lookout station. One of the principal regiments to serve on the island was the First Georgia Regulars. Duty on the island was far from enjoyable as the troops were plagued by isolation, boredom, and swarms of blood-sucking mosquitoes and sand flies.

After the Confederate loss of Fort Walker and Fort Beauregard on nearby Port Royal Sound, South Carolina, it was realized that the South could not hold such isolated positions as barrier islands without a strong navy. On November 10, 1861, the Tybee Island defenses were abandoned. The Confederates blew up part of the lighthouse to keep it from being used by the Federals. A short time after the Confederate evacuation, Federal troops occupied Tybee Island.

The Union Army built a sand battery around the tower, and the tower and lighthouse would be in Federal hands until the end of the war. The lighthouse was soon operational again. Guns on Tybee Island were close enough to Fort Pulaski to cause serious damage, so the island would play a pivotal part in the bombardment of the fort on Cockspur Island in April of 1862.

After the war the tower was used as a post office and residence of postmistress Elizabeth L. Wortham. In 1887, the town was incorporated under the name of Savannah Beach.

After Fort Screven was built on the island in 1897, the new Fort Screven Post Office was housed in the tower. Mrs. Alice F. Evans was postmistress. The practice firing of the nearby big guns would rattle Mrs. Evans windows and empty her shelves of China. Because of this, she would be given an advanced warning before the guns were fired. The tower and its post office were gutted by fire on March 19, 1913. In 1914, the tower was razed to clear the field of fire for the big guns of the nearby batteries. The tower had stood for nearly a century. The site of the tower is now in the Atlantic Ocean due to beach erosion,.

Fort Pulaski, Cockspur Island Begun 1829

Cockspur Island is one mile in length and less than a half mile in width. Lying near the mouth of the Savannah River, the island got its name from the shape of its treacherous reef that points toward the open sound of Tybee Roads. Ever since colonial times this island, located only a couple of feet above sea level, has been considered to be strategic, vital to the defense of Savannah. Starting in 1829, Fort Pulaski was built on this island. Surrounded by a wet ditch, the fort is a large brick five-sided truncated hexagon with a large ravelin, called the demilune.

Fort Pulaski breach 1862 John Weaver Collection

Fort Pulaski is actually the third fort to occupy Cockspur Island. In 1761, because of Spanish incursions from nearby Florida, the British built a log blockhouse surrounded by a palisade and named it Fort George. In 1776, the fort was dismantled and abandoned by the patriots. Begun in 1794, the First System work named Fort Greene was constructed of earth and timber with a guardhouse and enclosed by a palisade. The fort was named for General Nathanael Greene, who resided at Mulberry Grove Plantation near Savannah. In 1804, a hurricane swept over the island and destroyed the fort, drowning part of the garrison.

When laying out the trace of Fort Pulaski, the Army Corps of Engineers positioned the salient of the truncated hexagon in such a way that the faces and flanks gave maximum coverage of the channels on each side of the island. This fort is massive and was constructed with an estimated 25 million bricks. The scarp was at least 7 ½ feet thick. There are 12 vaulted casemates in each pair of faces that meet at the center salient. There are 12 vaulted casemates in each flank curtain. Each casemate contained a single seacoast gun embrasure. A continuous vaulted gallery runs through all the

Fort Pulaski from the air John Weaver Collection

The salient of the demilune, demilune wet ditch, and two demibastions Photo by John Weaver

Gun emplacements on the terreplein of Fort Pulaski Photo by John Weaver

Casemate gallery shows the thick, massive vaulting Photo by John Weaver

casemates in the four seacoast fronts. The arches and support columns are massive. A pan coupe·is located in each angle where the flanks and faces meet at a salient. Each of the three pan coupe had one gun embrasure, as each one is only one casemate wide. The entire work is surrounded by a wet ditch. A canal from the river filled the ditch and was controlled by sluice gates.

A demibastion is located at the junction of the gorge wall and each flank wall. These demibastions were for flanking fire along the gorge, sallyport, and the demilune counterscarp. A flank howitzer embrasure and smaller musket loopholes are in the gorge curtain angles as well in the faces of the demibastion, allowing fire down the ditch in front of the demilune.

The gorge contained vaulted quarters for the garrison. In the gorge curtain are musket loopholes that also served as windows for the quarters. On the parade side of the gorge each of the quarters had a door and two windows. A veranda ran most

The gorge quarters from the parade. In the center is the sallyport
Photo by John Weaver

A corner where the casemates meet one of the gorge magazines and quarters
Photo by John Weaver

of the length of the gorge. On each end of the gorge are magazines and stairs leading to the barbette tier. A brick blindage was built in front of each magazine to protect them from direct fire across the parade. In the center of the gorge is the sallyport with a drawbridge connecting the sallyport to the ravelin. In the sallyport was a set of vertical and horizontal bars, called a portcullis, that could be lowered to bar entry. This drawbridge would raise and the portcullis would lower synchronously. Inside the portcullis, heavy double doors in the sallyport also barred entry. Musket loopholes lined the sallyport on both sides. Any enemy caught in the sallyport could be fired on from these loopholes.

Closeup of the two demibastions and gorge curtain with the sallyport
Photo by John Weaver

The pan coupe that was breached during the bombardment
Photo by John Weaver

For land-side defense, a very large triangular ravelin, known as the demilune (half moon), was built to protect the gorge curtain and provide cannon facing the landward approach to the fort. The demilune is surrounded by the wet ditch. A drawbridge gave entrance across the demilune ditch to the demilune and another drawbridge gave entrance from the demilune across the ditch to the sallyport. The triangular demilune mounted guns en barbette. A large hot-shot furnace was once on the demilune.

The fort's large parade has an area of 2½ acres. Two large hot-shot furnaces were once on the parade. The terreplein of each face, flank, and gorge that surrounds the parade mounted guns en barbette. The barbette tier of each demibastion mounted guns en barbette to fire along the gorge curtain and the demilune ditch.

Although the massive 25-foot walls are impressive, even more unseen material lies below the fort's foundation. To distribute the weight of the massive walls on the soft wet soil of the site, a system of reverse arches was used to transfer the weight of the walls evenly to a timber grillage below the floor. The base of the outside wall is splayed out to distribute the weight even more. In order to reduce weight, the floors of the casemates were constructed of wood.

Although the seacoast fronts have no flanking structures, an early plan shows a caponier in the ditch of the southeast pan coupe. Curiously, there is only one caponier shown on the plan. The caponier was to have been constructed on pilings in the ditch. Musket loopholes on each side of the caponier would open along the curtains. The caponier was to have been entered by a ladder suspended from the pan coupe gun embrasure.

Smoothbore cannon and embrasure in a massive vaulted casemate
Photo by John Weaver

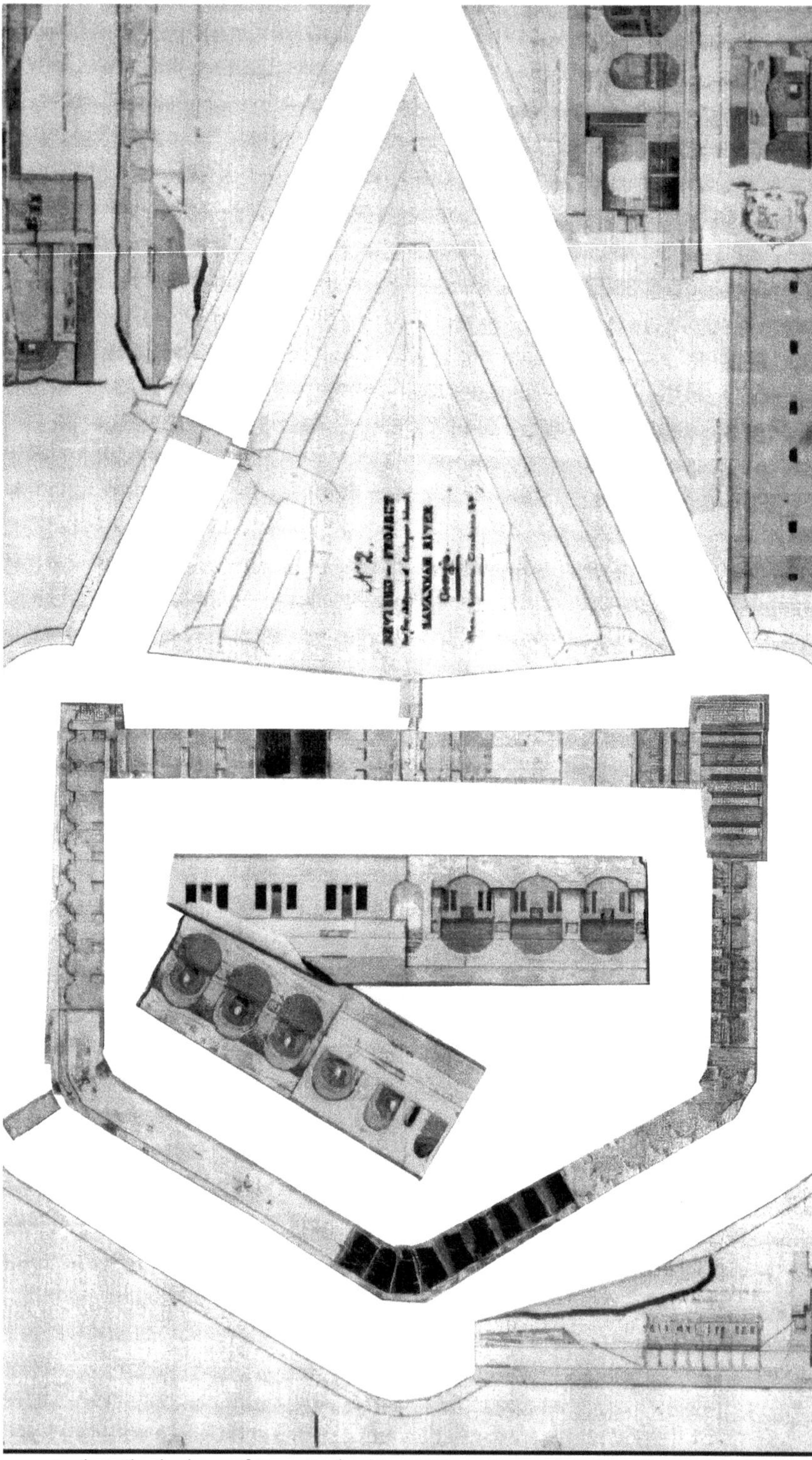

A detailed plan of Fort Pulaski and its triangular demilune. NARA

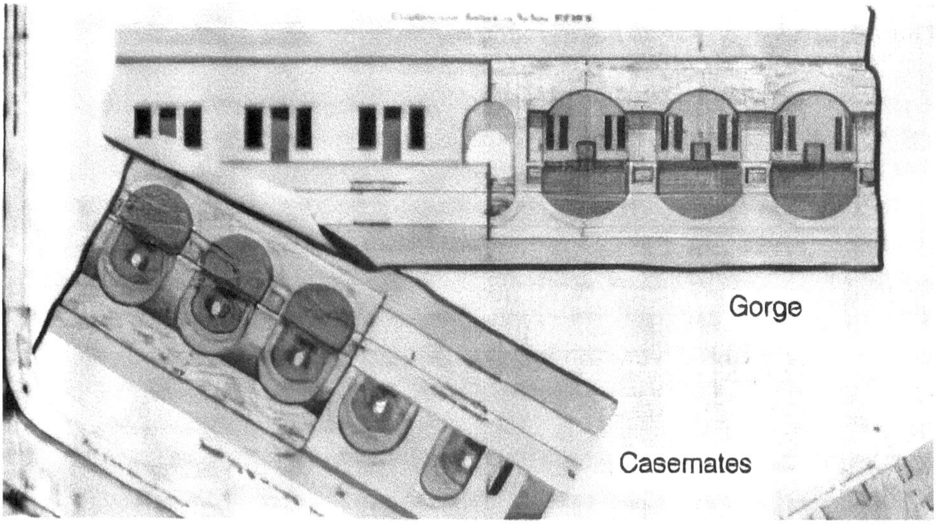

Details of the inverted arches of the foundation, casemates, and gorge rooms NA

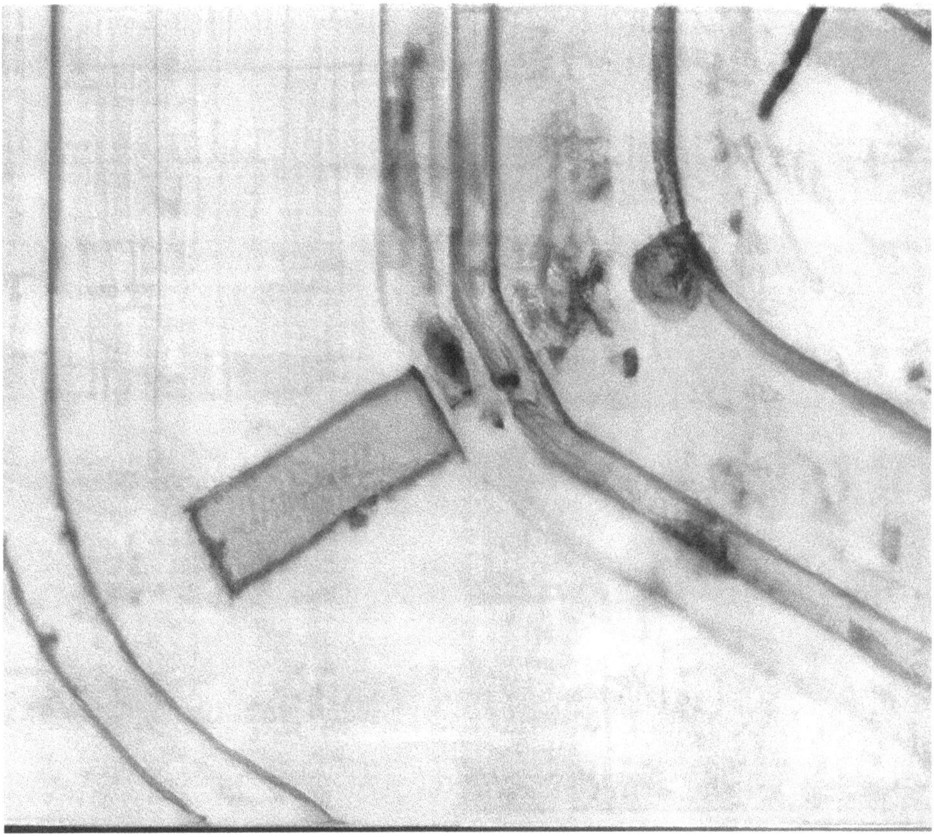

Detail of the caponier that was proposed for the southeast pan coupe NA

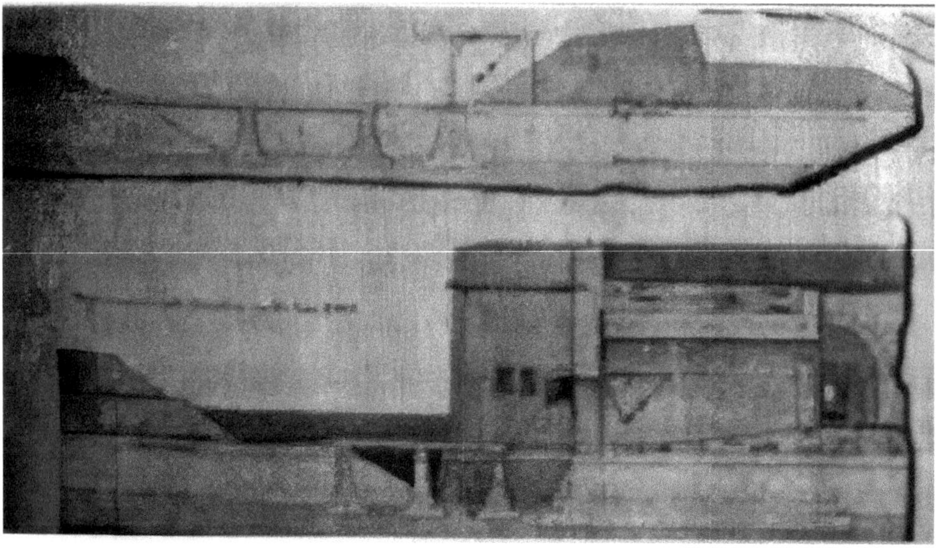

Section of the ramparts and wet ditch of Fort Pulaski Dale Manuel Collection

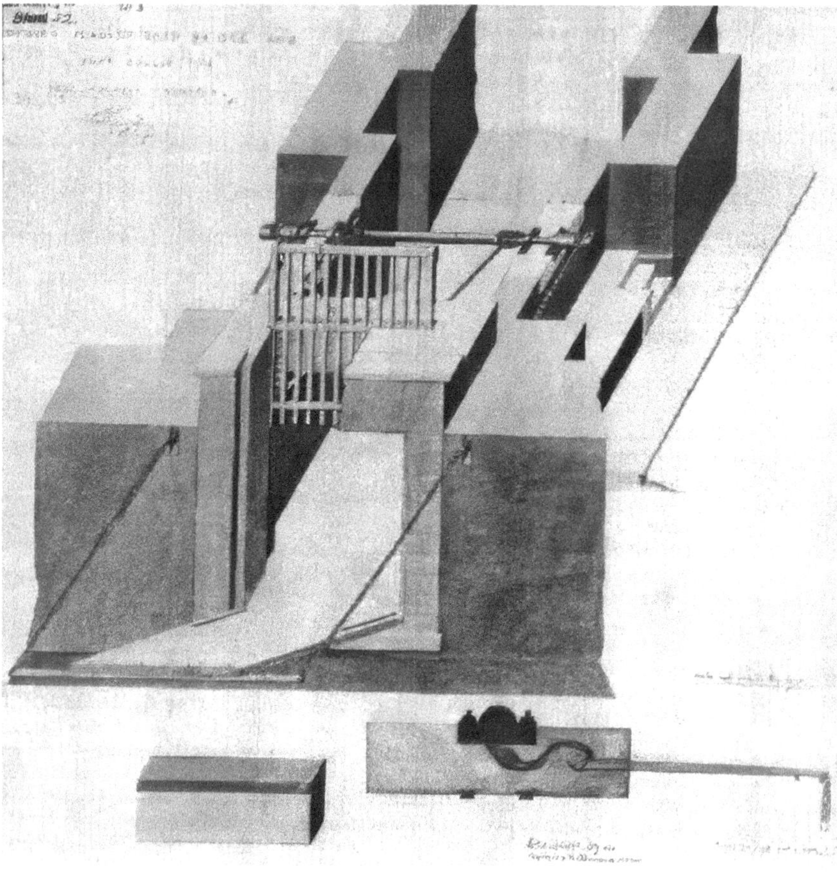

Isometric view of the drawbridge and portcullis Dale Manuel Collection

Section of a casemate and an embrasure Dale Manuel Collection

Section of the sallyport and drawbridge and portcullis mechanism. When the drawbridge was raised the portcullis would lower to bar entry. NA

Part of the drawbridge and portcullis mechanism Dale Manuel Collection

The drawbridge that was over the demilune wet ditch. The mound on the right once held the mining casemate for the North Channel minefield during the Endicott Period Dale Manuel Collection

A rain collection system on the terreplein piped water to cisterns below the floors. Ten cisterns in all provided as much as 200,000 gallons of fresh water for the garrison.

The fort was designed by General Simon Bernard in the finest traditions of polygonal fort design. The original plans called for three tiers, but the weight of the masonry would have been too great for Cockspur Island's soft, wet soil to support. The plan was modified for two tiers with its ingenious reversed-arch foundation system. The first supervising engineer of the project was Major Samuel Babcock, who started construction in December of 1828. The dikes, canals, and the trace of the fort were laid out by young Lieutenant Robert E. Lee. This was Lee's first engineering assignment after graduating from the U. S. Military Academy at West Point in 1829. Because of Babcock's ill health, Lieutenant Joseph Mansfield became supervising engineer and is considered the builder of Fort Pulaski. Mansfield served until 1845, when he was called away to Texas. Although the fort was considered finished in 1847, by 1860 only 20 guns of the planned 146 guns were mounted.

In 1833, the fort was named in honor of Polish Count and American General Casimir Pulaski. On October 9, 1779, Pulaski was mortally wounded while leading a charge on the British defenses of Savannah. While being taken off the field he gave his last order to Light Horse Harry Lee: "Follow my lancers, to whom I have given my order of attack." On board the Wasp, he died two days later and was buried at sea near the mouth of the Savannah River.

On January 3, 1861, State of Georgia forces under Colonel Alexander R. Lawton seized Fort Pulaski from a single ordnance sergeant. The troops were ordered to seize the fort by Georgia Governor Joseph E. Brown to prevent it from being garrisoned by Federal troops. In October, 1861, Major Charles H. Olmstead was assigned command of Fort Pulaski. In November, 1861, nearby Tybee Island was abandoned. This decision to abandon the position of Tybee would be crucial in the upcoming bombardment and reduction of Fort Pulaski.

General Robert E. Lee inspected the fort in January, 1862, and told Colonel Olmstead, pointing at the western end of Tybee Island, "They will make it very warm for you with shells from that point, but they cannot breach your walls at that distance." Lee, like most of the military leaders in this nation, was unaware of the range of the new rifled guns and the distance from which they could damage masonry. Chief of U.S. Army Engineers Joseph G. Totten proclaimed that "You might as well bombard the Rocky Mountains" as bombard Pulaski. One U.S. officer who was aware of recent experiments with rifles and masonry that were conducted in England and Prussia was Army engineer Captain Quincy A. Gillmore. Soon to be a General, Gillmore believed that rifled artillery might be successful in breaching masonry forts at distances too great for smoothbore guns. Starting on February 19, 1862, General Gillmore began building eleven batteries that contained a total of thirty-six rifles, Columbiads, and heavy mortars along the northwest shore of Tybee Island. The batteries contained

Joseph Mansfield was considered the builder of Fort Pulaski. He supervised construction from 1831 to 1845. He became a Union General during the Civil War and was killed in action at Antietam Dale Manuel Collection

Drawing of the bombardment of Fort Pulaski from the Tybee batteries Dale Manuel Collection

General Quincy Gillmore who was successful in the reduction of Fort Pulaski
Library of Congress

Interior of a demibastion Photo by John Weaver

magazines and bombproofs. The distances from the batteries to the fort ranged from 1,650 yards to 3,400 yards. In Batteries Sigel and McClellan, two of the most advanced batteries on Goat Point, there were ten rifled guns. The batteries contained five 30-pounder Parrott rifles, one 24-pounder, two 42-pounder and two 32-pounder James rifles. Although called James rifles, the artillery was actually old 24-, 32- and 42-pounder guns that had been sleeved to fit a James projectile. The rifled James projectiles were double the weight of the old round projectiles fired by the original guns. The James projectiles weighed 48, 64, and 84 pounds, respectively.

In February, 1862, Federal troops built six-gun Battery Vulcan across the river at Venus Point on Jones Island. The Federals also built six-gun Battery Hamilton on

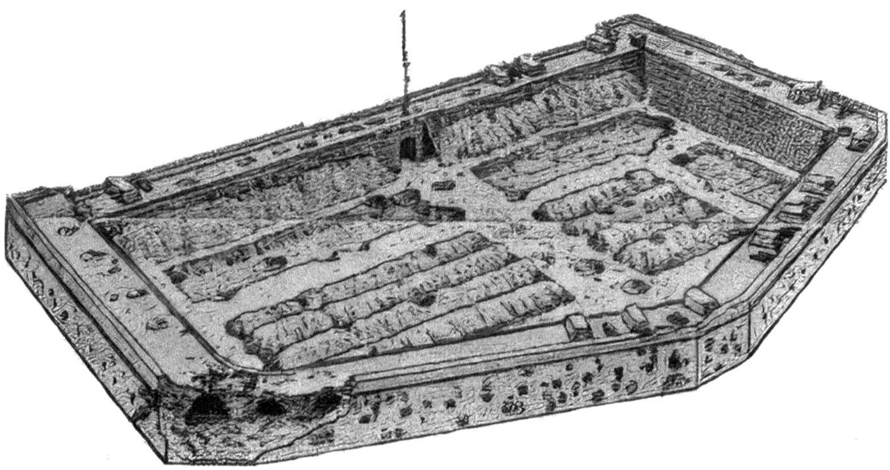

Drawing of the damage done to the fort. The Confederates shielded the casemates with a blindage made of lumber and earth Dale Manuel Collection

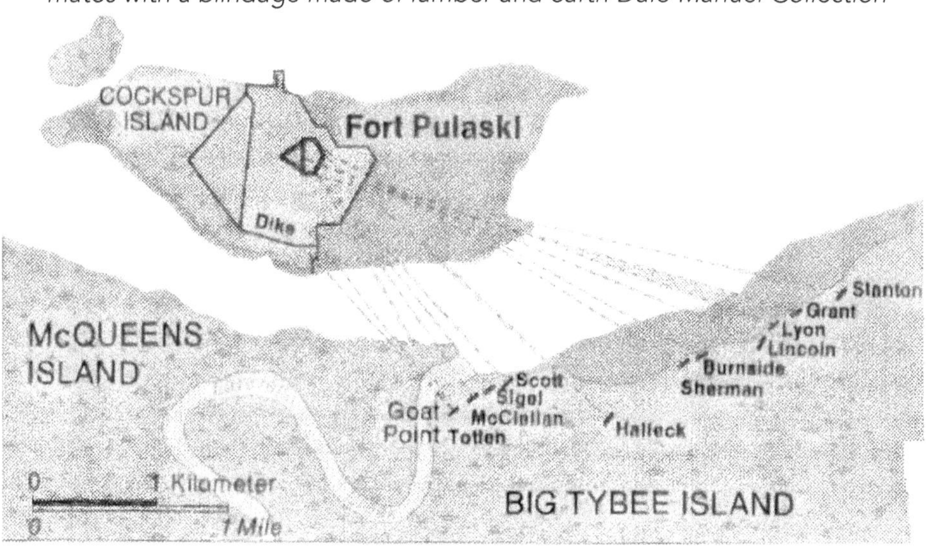

Map of the positions of the Federal batteries that fired on Fort Pulaski Dale Manuel Collection

A projectile is still buried in the damaged scarp of the fort
Photo by John Weaver

the western end of Bird Island. With the obstruction of Lazaretto Creek, the Federals succeeded in isolating Fort Pulaski.

Inside Fort Pulaski there were 45 guns and outside the fort were three mortars in batteries. The Confederate garrison consisted of 24 officers and 337 men.

On the second day of bombardment the guns on the fort's ramparts were dismounted and the southeast pan coupe was breached. Explosive shells threatened the powder magazine in the northwest corner of the fort. Hopelessly cut off from help and concerned about the lives of his men, Colonel Olmstead surrendered after only 30 hours of bombardment. Approximately 5,275 projectiles were hurled at the fort and its beleaguered garrison.

The captured Confederates were imprisoned at Governor's Island in New York Harbor. Colonel Olmstead would later be exchanged. The fort would be in Federal hands through the end of the war and served as a prisoner-of-war camp. After the war, the fort housed political prisoners.

In the early 1870s the demilune received 15-inch Rodman gun emplacements and four covered magazines. The fort was on a caretaker status into the 1880s when it was abandoned by the caretaker and lighthouse keeper. During the Spanish-American War, the fort was reactivated and the 15-inch Rodman battery was repaired and put into service. During the Endicott period a magazine in the demilune was converted to a mine casemate that controlled the north channel minefield. Cable tanks were installed under two of the brick casemates of the old fort to hold mine-cable reels. Construction began in June, 1899, on Battery Hambright near the river to guard the minefields. The battery mounted 3-inch rapid-fire steel breechloading rifles. The bat-

Battery Hambright from the back Mark Berhow Collection

3-inch gun emplacement at Battery Hambright Mark Berhow Collection

tery was named in honor of Horace George Hambright, a young officer who had died on the Dakota frontier in 1896.

The fort was declared a National Monument in 1924. Restoration work by the CCC began about 1933. In WW II the island was used as a Navy and Coast Guard Station. On Pulaski Day, October 11, 1947, the 168th anniversary of the Count's death, the fort was re-opened as a National Monument.

Rodman emplacements in the demilune Dale Manuel Collection

Confederate Batteries, Near Savannah 1861-1864

Drawing of men working on a Confederate gun battery Dale Manuel Collectio

Early in the Civil War, the Confederates built earthwork batteries on the barrier islands close to the Atlantic to guard Savannah. When these isolated outposts proved to be untenable, they moved their defenses further inland to guard the principal waterways and coastal railway links that led to the city. Taking advantage of the natural defensive barriers of marshland, a line of batteries ran from the Great Ogeechee River to the Savannah River and on to Charleston, South Carolina. Savannah and most of the batteries would be evacuated on December 20, 1864.

In 1861, the barrier islands of Tybee and Wassaw were fortified with batteries. Tybee had a two-gun battery erected near the old Martello Tower to guard the entrance to the Savannah River and to prevent landings on the island. The battery on

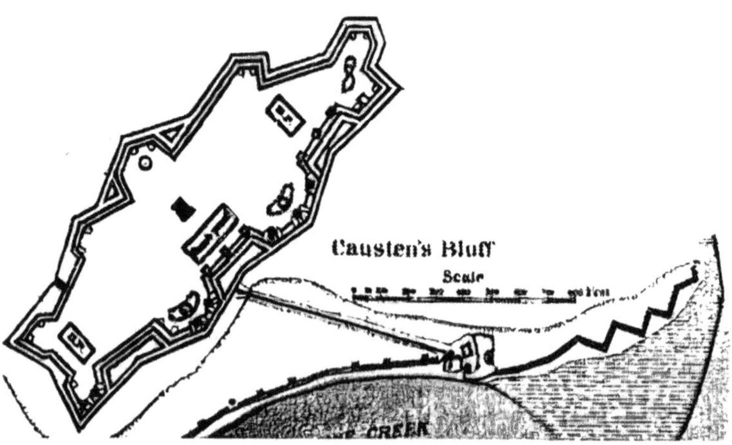

Fort Bartow on Caustern's Bluff guarded St. Augustine Creek and the Savannah River Dale Manuel Collection

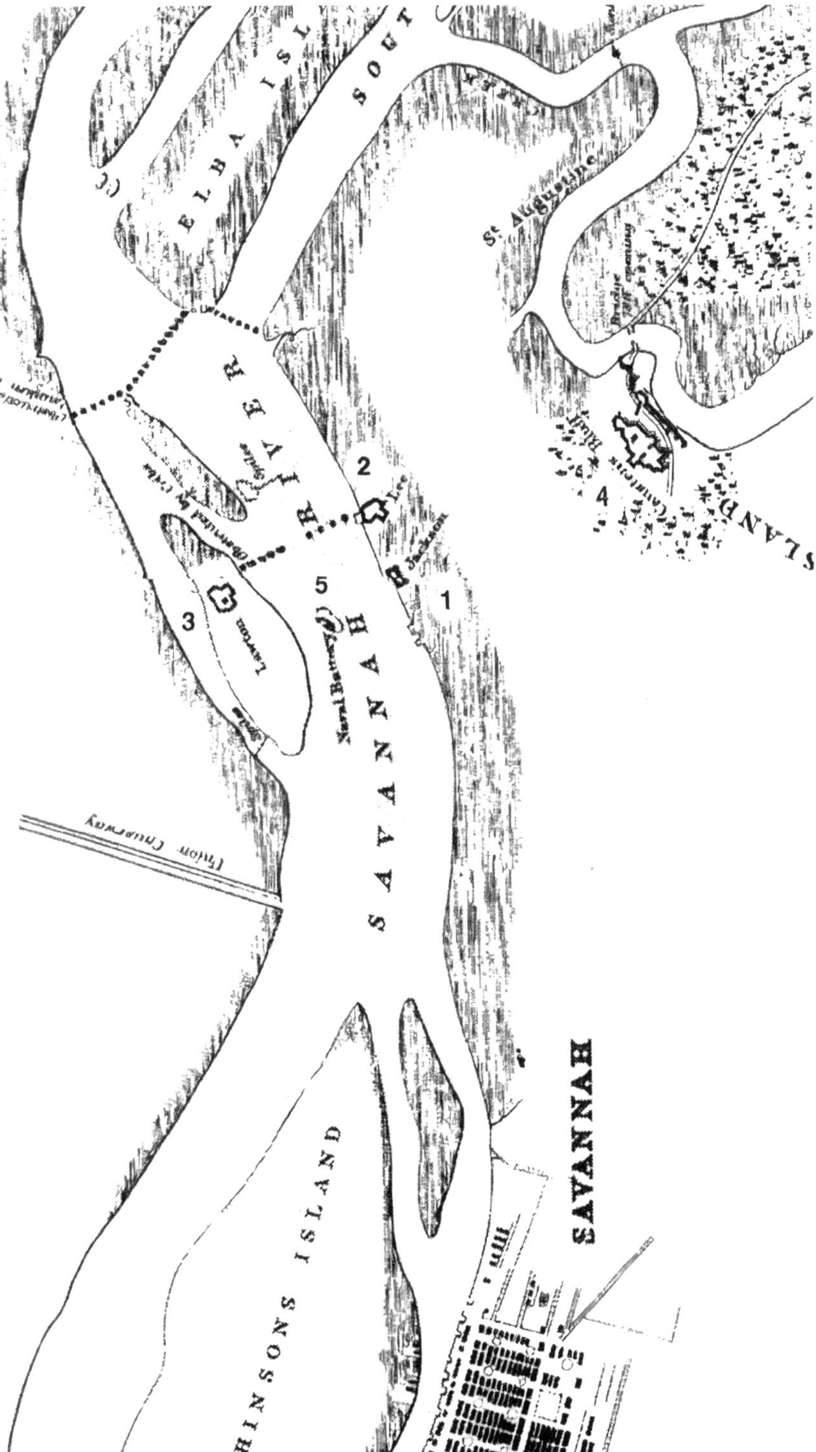

Legend: 1. Fort Jackson; 2. Battery Lee; 3. Fort Lawton; 4. Fort Bartow; 5. Battery Tattnall Dale Manuel Collection

Wassaw was erected to guard the entrance to Wassaw Sound. A ten-gun battery was erected on Green Island on Ossabaw Sound to guard the mouth of the Vernon River. A battery was also erected on Skidaway Island at the mouth of the Wilmington River. By November, 1861, the Tybee and Wassaw Island Batteries had been evacuated. The decision to abandon the position on Tybee Island would prove to be crucial in the bombardment and reduction of Fort Pulaski on April 11, 1862. In March, 1862, the works on Skidaway and Green Island were abandoned.

After the fall of Fort Pulaski, old batteries were strengthened and new batteries were built. The Savannah River approach was strengthened with the construction of Batteries Lee and Tattnall, and Fort Lawton. Battery Lee was built adjacent to Fort Jackson for increased firepower on the river. Battery Tattnall, sometimes called the naval battery, was built on a small island near the middle of the channel. In order to create a crossfire with Battery Lee and Fort Jackson, Fort Lawton was built on Harnwell Island on the South Carolina side of the river. Pilings and submerged cribbing blocked the main channel.

Other than the Savannah River, the Wilmington River through St. Augustine Creek was the most direct route to the city. Fort Thunderbolt was a large earthwork that guarded the Wilmington River from enemy attack from Wassaw Sound. Other batteries along St. Augustine Creek included Battery Bonaventure and Fort Bartow at Caustern's Bluff. Covering about seventeen acres, Fort Bartow was the largest earthwork fort. It not only guarded the St. Augustine Creek approach but was close enough to the Savannah River to guard it as well. The fort's armament consisted of two 10-inch mortars, two 8-inch columbiads, and one 32- pounder gun.

On Whitemarsh Island, at Turner's Rock, a strong battery was constructed where Turner's Creek emptied into the Wilmington River. A line of earthworks containing four bastions for flanking fire stretched across Whitemarsh Island and a small battery was built on the east side to guard the Wilmington Narrows. There were several skir-

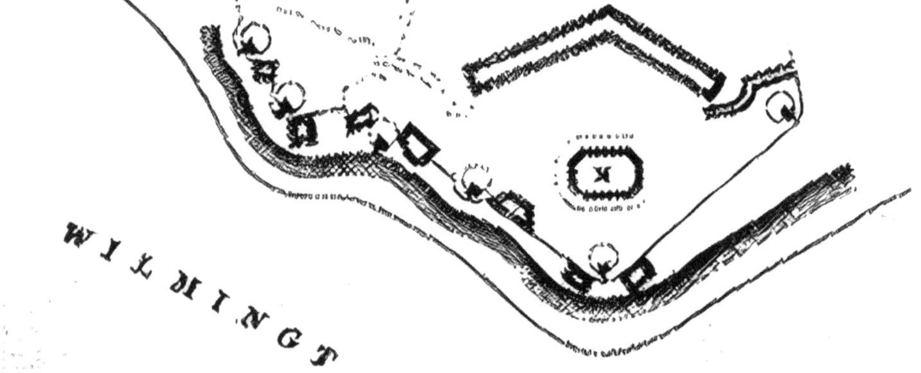

Turner's Rock Battery was located on the southern end of White Marsh Island. It guarded the Wilmington River at the mouth of Turner's Creek
Dale Manuel Collection

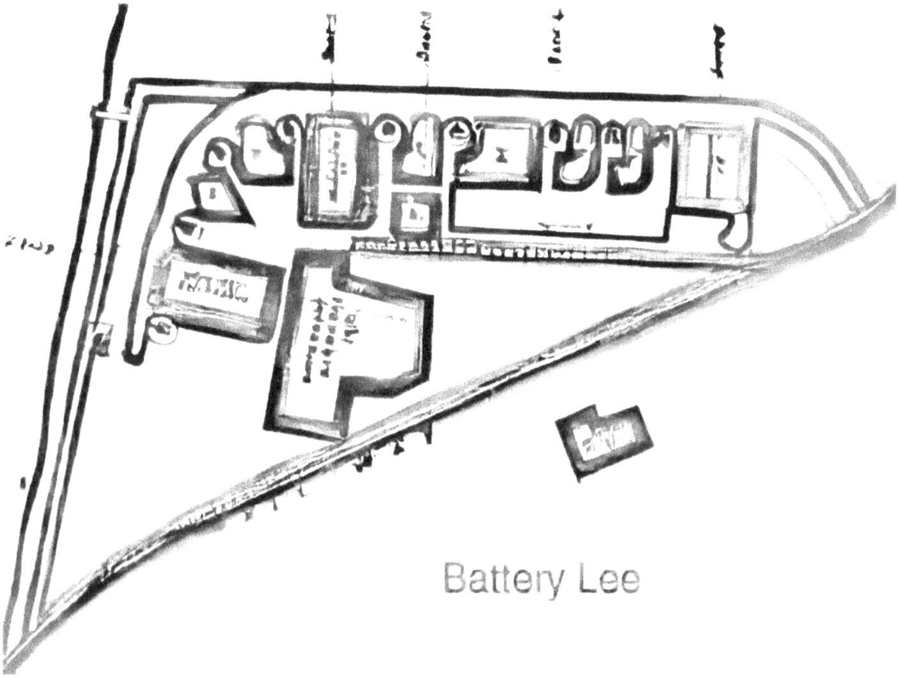

Battery Lee Dale Manuel Collection

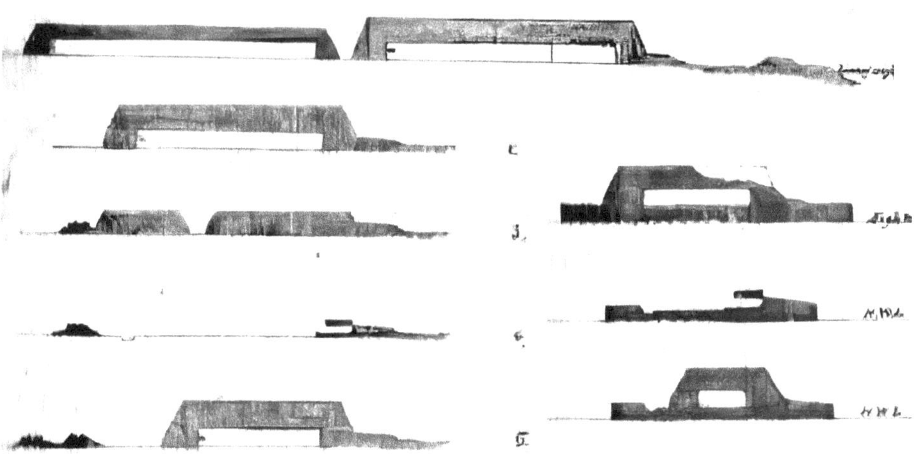

Profiles a, b, c of bombproof, magazines, and traverses of Battery Lee. Dale Manuel Collection

mishes on March 30 and 31 and again on April 16, 1862. On February 20, 1864, a Federal landing party advanced to the bridge at Oatland Island, but were repelled and retreated to their boats.

Other batteries were built on Rose Dhu Island, Beaulieu, and the Isle of Hope. Battery Beaulieu guarded the Vernon River. Fort Wimberly was constructed on the Burnside River. Battery Daniels was built at the forge on the Isle of Hope.

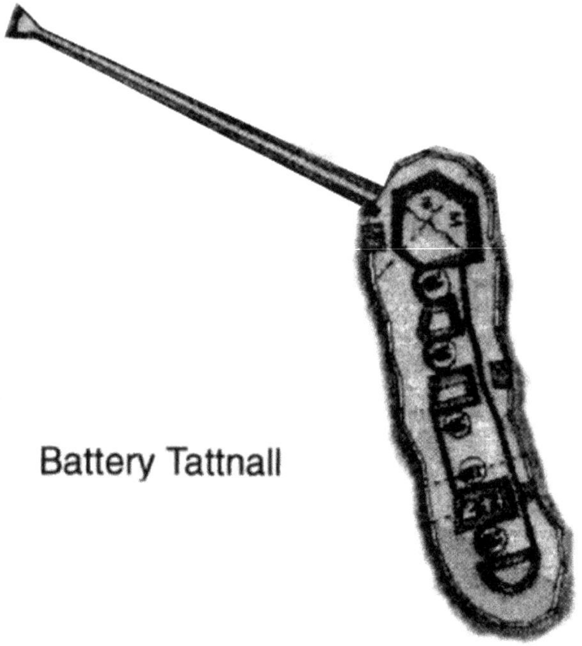

Plan of Battery Tattnall that was erected in the channel between Fort Jackson and Fort Lawton. A long wharf supplied the garrison Dale Manuel Collection

A semicircular line of earthworks guarded the land approaches to Savannah. Fort Boggs on Brewton Hill was one of the star-shaped strongholds of these defenses, designed by Confederate engineer Captain William R. Boggs, who was later a Confederate General. Another defense in the line was Fort Brown near the Catholic Cemetery.

Fort McAllister, Great Ogeechee River Begun 1861

In June, 1861, construction began on a Confederate earthwork twelve miles south of Savannah on the south bank of the Great Ogeechee River. The fort was to guard the mouth of the river and the Atlantic and Gulf Railroad bridge over the Ogeechee, a few miles to the north. It was named for Confederate Lieutenant Colonel Joseph L. McAllister on whose land the fort was built. Although this was one of the most isolated defensive outposts of Savannah it was, perhaps, the most important. Officers on both sides saw it as the key to Savannah because it guarded the back door to the city. Throughout the war the fort's defenses would be tested many times and it would come to symbolize Rebel defiance.

On July 29, 1862, the first of many attacks occurred as four Union gunboats, the *Paul Jones, Unadilla, Huron* and *Madgie,* engaged the fort. The gunboats, under the command of Commander Charles Steedman, did little damage. On November 19, 1862, the fort was bombarded for five hours. On January 27, 1863, there was an artillery duel with the monitor *Montauk* that had come close to the fort firing 15-inch shells from her two Dahlgren guns. On February 1, 1863, the *Montauk* dueled with the fort for five hours. The Union monitor was hit forty-six times, but the Confeder-

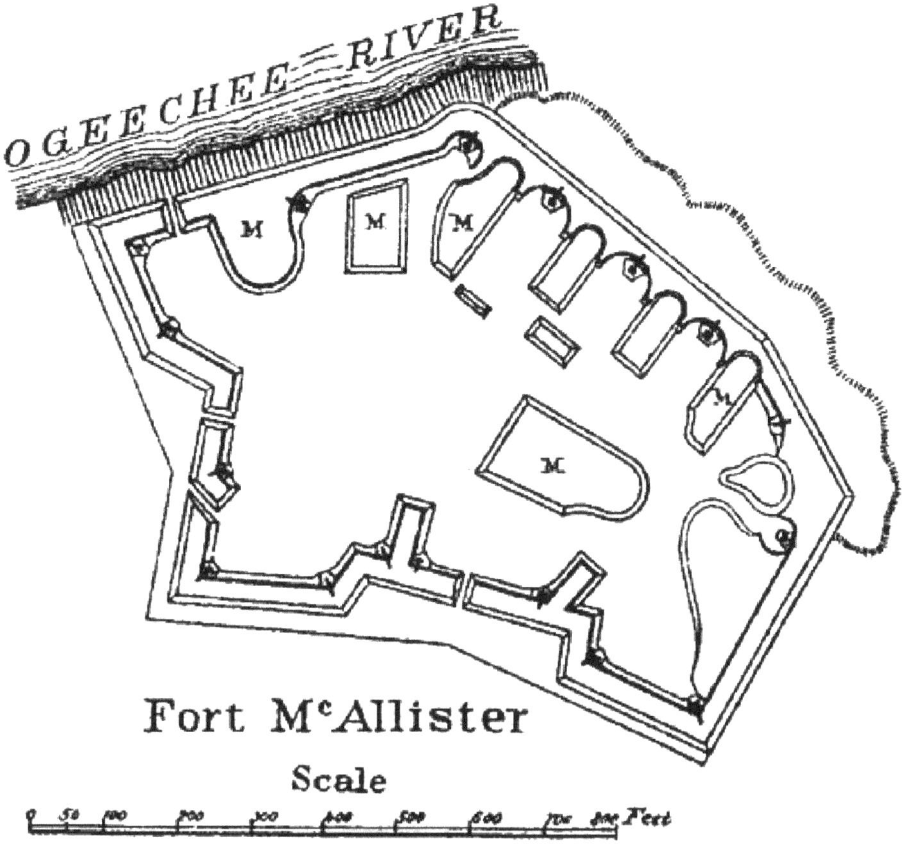

Plan of Fort McAllister on the Ogeechee River. This plan was prepared by Federal engineers in 1864. Shown are five magazines with most of the guns aimed toward the river Dale Manuel Collection

USS Montauk attacks Fort McAllister Harpers Weekly

Fort McAllister after capture by Union forces
Photographic History of the Civil War

ates lost their commander, Major John B. Gallie, who was killed when the 32-pounder he was commanding was hit and disabled.

The new commander of the 230 Confederates would be Major George W. Anderson, Jr. The armament at the fort consisted of one 10-inch mortar, three 10-inch Columbiads, one 42- and one 32-pounder rifle, four 32- and one 24-pounder, two 12-pounder mountain howitzers, two 12- and six 6-pounder field pieces.

On February 28, the *Montauk* and two gunboats attacked again. This time the blockade runner *Rattlesnake,* better known as the raider *Nashville,* had become stranded slightly upriver from the fort while trying to escape to the Atlantic Ocean. The *Montauk* bombarded the helpless ship and set it on fire. Its crew abandoned the doomed *Nashville.* The *Montauk* then hit a torpedo (mine) in the river while returning. The *Montauk's* commander John L. Worden ordered the monitor run aground to prevent it sinking. The hull was patched, and it was towed to Port Royal Sound for repairs.

On March 3, Fort McAllister sustained a seven-hour bombardment by the three monitors: the *Passaic, Patapsco,* and *Nahant.* The fort's only fatality was the garrison's mascot "Tom Cat." The Union commander of the monitors, Captain Percival Drayton, summed up the damage to the fort by saying, "No injury was done which a good night's work would not repair." Northern Journalist Horace Greeley remarked that the Union sailors "saved their ammunition by letting McAllister alone."

The biggest threat to the fort would not come from the sea but from the land with the arrival of General Sherman's "March to the Sea." The fort was never intended to resist a strong siege. The attacks started on December 12, 1864, when cavalry troops under General H.J. Kilpatrick skirmished with Rebel pickets. With McAllister's big guns facing the sea approach, 4,300 troops under Brigadier General William Hazen crossed the rebuilt King's Bridge and massed for an attack on the rear of the fort. In preparation for a land attack, the Rebels cleared a field of fire, dug a deep ditch, constructed an abatis, and planted primitive land mines to the rear of the fort.

The main attack commenced after 4 P.M. on December 13, 1864. The small garrison was overwhelmed and fought hand-to-hand with the Federals coming over the ramparts. Many Confederates scrambled into bombproofs and fought there at close quarters. The assault only lasted about fifteen minutes and by 4:30 P.M. the fort was in Union hands. The Confederate losses were 14 dead, 21 wounded and 195 captured. The Union casualties were 24 dead and 110 wounded. Most of the Union losses were due to the Rebel mines. After the battle, details of captured Confederate prisoners were forced to dig up the remaining mines.

The fall of Fort McAllister meant the Union Army could now be supplied from King's Ferry, allowing the Navy to bring up heavy rifled guns and ammunition for the reduction of Savannah. The back door to the city was now open. Confederate General Hardee was compelled to evacuate Savannah.

The Georgia Navy

The first Georgia State Navy organized to guard Savannah began in 1861. The first squadron was a collection of steamers and tugs converted into gun boats and known as the Mosquito Fleet. The small navy was commanded by Commodore Josiah Tuttnall, a naval veteran of the War of 1812, the Mexican War, and the West Indian and Far Eastern Squadrons. 'l'uttnall was the second captain of the CSS *Virginia,* which he ordered destroyed to prevent her capture when her deep draft prevented escape from Norfolk, Virginia.

Commodore Josiah Tattnall was the son of a Georgia governor and long-time naval veteran. He helped salvage the British admiral's barge the Centipede after the Battle of Cranney Island during the War of 1812. He was wounded during the bombardment of the Veracruz forts in the Mexican War. He coined the phrase "blood is thicker than water" to explain why a neutral had aided a British and French attack on Chinese forts on the Pei-ho River in 1859 Dale Manuel Collection

The first flagship and the first vessel to bear the city's name was the gunboat GS *Savannah*. Originally the paddlewheel river steamer the *Everglade,* the *Savannah* mounted a single 32-pounder gun. Other vessels of the Mosquitoes were two converted tenders, the *Sampson* and the *Resolute,* that mounted one 32- and one 12-pounder each. Two other vessels obtained from Charleston were the *Lady Davis* and the side-wheeler *Huntress.*

On November 4, 5, and 7, 1861, Tattnall boldly attacked a vastly superior Federal Fleet during action at Port Royal Sound. After firing from long range, the out-gunned Confederates retreated to Skull Creek before they were eventually annihilated. The *Savannah* was commanded by Lieutenant Maffitt, who would later captain the raider the CSS *Florida.* The *Sampson* was skippered by Lieutenant J.S. Kennard and the *Resolute* by Lieutenant J. Pembroke Jones. The flotilla attacked again on November 5 but was quickly driven off. The *Savannah* was damaged that same day when Maffitt came too close to the Federal gunboats. Maffitt would later receive a reprimand for his rashness. With the loss of Forts Walker and Beauregard that had been bombarded by DuPont's Federal Fleet and subsequently overrun by amphibious assault on November 7, the Union Navy would have a deepwater port on Southern soil to blockade the South Atlantic Coast. The Mosquitoes limped back to Savannah.

In the fall of 1862, the Tift brothers (Henry, Willink, and Krenson) and Hawkers Shipyards were busy constructing ironclads. The first ironclad to be constructed was the CSS *Georgia.* Originally known as "The Ladies Gunboat" because its building was funded by The Ladies Gunboat Association, the *Georgia* was 250 feet long and was armored with railroad iron. Because the *Georgia's* engines were underpowered and she leaked so badly, she was used instead as a floating battery. The *Georgia* was moored near Elba Island where she could be turned to cover both channels.

The ironclad CSS *Atlanta* was commissioned on November 22, 1862. The *Atlanta* had been converted from the iron-hulled steamer the *Fingal.* At over a thousand tons and 204 feet in length, she was a formidable warship. Her English engines could propel her to a top speed of six knots. The *Atlanta's* armament consisted of two seven- and two six-inch Brooke guns. On the bow was an iron ram and a spar torpedo.

The ironclad CSS *Savannah* was commissioned on June 30, 1863. It was 150 feet in length and had a crew of 27 officers and 154 men. Her armament was four 6.4-inch Brooke rifles in addition to a spar torpedo and a ram. Steering problems, deep drafts, leakage, and poor ventilation plagued the ironclads.

The Atlanta after her capture at Wassaw Sound Dale Manuel Collection

The Atlanta is shown in dry dock after her capture

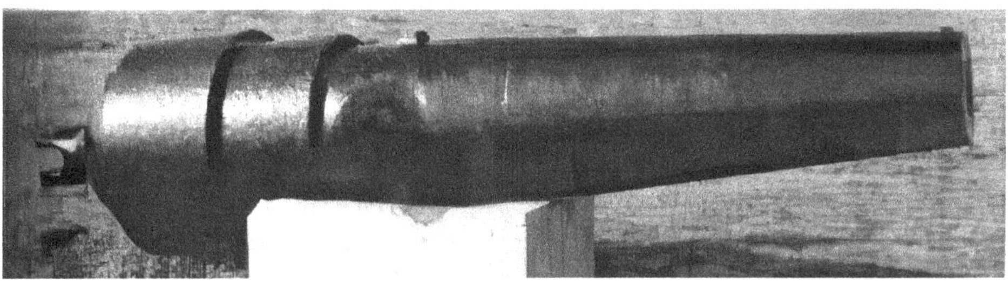

The CSS Atlanta had two of these Confederate-cast 7-inch Brooke rifles. These heavy rifles were considered the best cast by either side

On January 17, 1863, the CSS *Atlanta,* commanded by Lt. William Webb, ran aground on a sandbar in Wassaw Sound while attempting to engage the two monitors, the USS *Weehawken* and the USS *Nahant.* The *Atlanta* listed so badly most of her guns could not be brought to bear on the monitors. The stranded *Atlanta* was forced to surrender after the *Weehawken's* two large Dahlgren guns pounded her from close range.

Other vessels built at Savannah were the *Isondiga,* the *Macon,* the unfinshed ironclad *Milledgeville,* and an unnamed and unfinished ironclad. The *Macon* was a wooden gunboat with steam and sails that had six guns. The wooden gunboat *Isondiga* was a stern wheeler of 112 feet in length with one 9-inch gun and one 11.4-inch rifle. The *Milledgeville* would have been the sister ship of the CSS *Savannah* had it been completed. Other vessels were used for transports, tenders, and tugs by the Confederates.

One of the most daring acts of the war occurred on the night of June 4, 1864, when the USS *Water Witch* was captured in Ossabaw Sound by a boarding party from seven Confederate boats. After hand-to-hand combat, the *Water Witch* was captured by the Rebels. The Confederate commander, Lieutenant Thomas P. Pelot, was killed in the action. Thewas taken to White Bluff to avoid recapture.

While planning the evacuation of Savannah, a plan was made for at least part of the flotilla to escape to sea, but the same obstructions and torpedoes that protected the

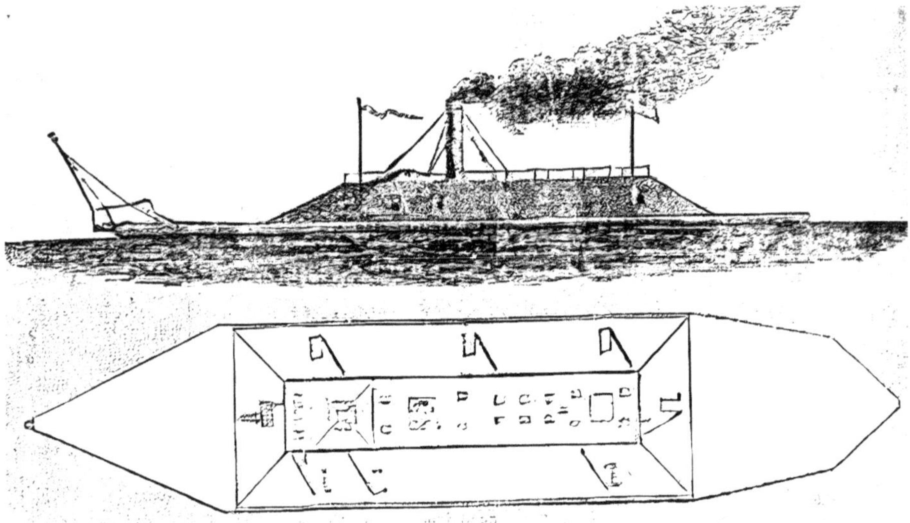

Elevation and plan of the Atlanta with its ironclad casemate and spar torpedo
Dale Manuel Collection

city made escape impossible. During the evacuation on December 20 and 21, 1864, most of the vessels of the flotilla would be destroyed to prevent their capture. The CSS Georgia would be scuttled by her crew near Fort Jackson. The CSS Savannah would serve as a rear guard for the retreating Confederates and was blown up by her crew on the night of December 21, to prevent capture.

Drawing of the destruction of the CSS Savannah. The ship was blown up by her crew to prevent capture by Union forces that besieged the city
Dale Manuel Collection

The Rebel Flotilla

Ironclads:

CSS *Georgia,* Commissioned July 1862, a floating battery, 9 guns, Dimensions 150' by 60', Armor railroad iron, crew of 200. Scuttled December 20, 1864.

CSS *Atlanta,* ex-*Fingal,* 3 screws, 1,006 tons, dimensions 204'x41'x15' 9", 8 knots, crew of 145, two 7" Brooke & two 6.4" Brooke, spar torpedo, Armor 4" iron & 5" iron deck, captured January 17, 1863, repaired and commissioned into U.S. Navy in 1864.

CSS *Savannah,* Commissioned June 30, 1863, Dimensions 173'x150'x12 one screw, speed 6 knots, two 7-inch Brooke & two 6.4-inch rifle, Armor 4" iron & 22" wood, crew-180, Destroyed December 21, 1864, to prevent Capture.

Milledgeville, sister ship of the Savannah, unfinished, burned.

Unnamed, unfinished ironclad burned on stocks Dec. 21, 1864.

Gunboats:

CSS *Isondiga,* Commissioned 1863, Stern-wheeler, one shell gun, One 6.4 rifle, crew 60, burned to prevent capture Dec. 21, 1864.

CSS *Macon,* Ex-*Ogeechee,* Commissioned August 3, 1864, Dimensions 150x 25x8, two screws, crew-120, four 32-pdr, one 32-pdr rifle & one 9", speed 12 knots. Captured at Augusta May 1865.

"The Mosquito Fleet":

GS *Savannah,* Ex-*Everglade,* wooden side-wheeler gunboat, one 32-pounder, first flagship, damaged at Port Royal, Refitted as blockade runner *Oconee* and foundered at sea in August, 1863.

Sampson, side-wheeler, tugboat converted to gunboat, one 32- pdr & one 12-pdr.

Huntress, side-wheeler, 3 guns, speed 16 knots.

Lady Davis, One screw, one 24- & one 12-pdr.

Resolute, side-wheeler, 322 tons, crew 35. Captured on Dec.12, 1864, in an attempt to burn a railroad bridge north of the city.

Other Vessels:

Ida, side-wheeler, transport, captured and burned by Union forces near Argyle Island on December 10, 1864.

Leesburg, transport 1862-65.

Talomico, side-wheeler, two guns, crew 20, transport, sunk 1863. Robert Habersham, side-wheeler, transport, exploded in1863. Beauregard, side-wheeler, transport, captured January, 1865.

Firefly, side-wheeler, one gun, tender, burned Dec 21, 1864.

General Lee, side-wheeler, transport.

Amazon, side-wheeler, transport.

St. Mary, steamer, patrol boat.

Bartow, steamer, tender.

St Johns, steamer, tender.

Columbus, steam tug.

Fort Screven, Tybee Island Begun 1897

In 1875, land was acquired for batteries of Rodman guns on the north end of Tybee Island. These batteries were never constructed, but in 1897 construction began on a wharf, buildings, and concrete batteries. The fort was originally called Fort Tybee, then Camp Graham, in honor of General Montrose Graham, commander of the Atlantic Coast defenses. In 1899, the name of the fort was officially changed to Fort Screven. It was named in honor of General James Screven, who was killed in action near Midway Church, Georgia, in 1778 during the Revolutionary War. The post on Tybee contains six concrete Endicott Period batteries.

Battery Hambright was constructed on nearby Cockspur Island to protect the north channel minefield. Another two-gun battery was constructed on nearby Wassaw Island to guard the entrance into Wassaw Sound and the mouth of the Wilmington

Battery Garland once held a 12-inch rifle and now serves as the Tybee Museum
Photo by Mark Berhow

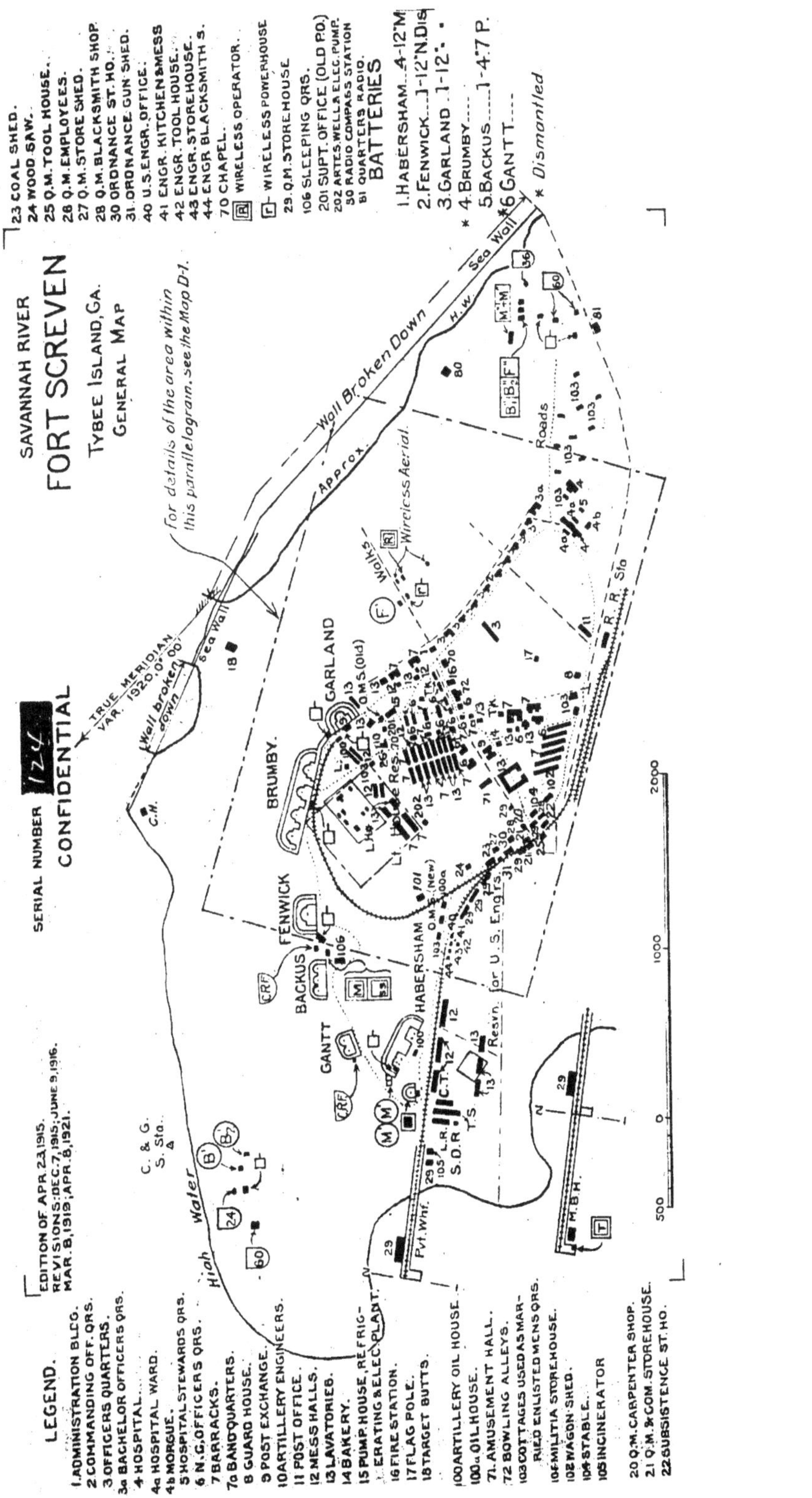
SERIAL NUMBER 124
CONFIDENTIAL
SAVANNAH RIVER
FORT SCREVEN
TYBEE ISLAND, GA.
GENERAL MAP
EDITION OF APR. 23, 1915.
REVISIONS: DEC. 7, 1915; JUNE 9, 1916.
MAR. 8, 1919; APR. 8, 1921.
TRUE MERIDIAN
VAR. 1920, 0°-00'
For details of the area within this parallelogram, see the Map D-1.
LEGEND.
1 ADMINISTRATION BLDG.
2 COMMANDING OFF. QRS.
3 OFFICERS QUARTERS.
3a BACHELOR OFFICERS QRS.
4 HOSPITAL.
4a HOSPITAL WARD.
4b MORGUE.
5 HOSPITAL STEWARDS QRS.
6 N.C. OFFICERS QRS.
7 BARRACKS.
7a BAND QUARTERS.
8 GUARD HOUSE.
9 POST EXCHANGE.
10 ARTILLERY ENGINEERS.
11 POST OFFICE.
12 MESS HALLS.
13 LAVATORIES.
14 BAKERY.
15 PUMP. HOUSE, REFRIGERATING & ELEC. PLANT.
16 FIRE STATION.
17 FLAG POLE.
18 TARGET BUTTS.
100 ARTILLERY OIL HOUSE.
100a OIL HOUSE.
71 AMUSEMENT HALL.
72 BOWLING ALLEYS.
103 COTTAGES USED AS MARRIED ENLISTED MENS QRS.
104 MILITIA STOREHOUSE.
102 WAGON SHED.
104 STABLE.
105 INCINERATOR
20 Q.M. CARPENTER SHOP.
21 Q.M. & COM. STOREHOUSE.
22 SUBSISTENCE ST. HO.
23 COAL SHED.
24 WOOD SAW.
25 Q.M. TOOL HOUSE.
26 Q.M. EMPLOYEES.
27 Q.M. STORE SHED.
28 Q.M. BLACKSMITH SHOP.
30 ORDNANCE ST. HO.
31 ORDNANCE GUN SHED.
40 U.S. ENGR. OFFICE.
41 ENGR. KITCHEN & MESS
42 ENGR. TOOL HOUSE.
43 ENGR. STOREHOUSE.
44 ENGR. BLACKSMITH S.
70 CHAPEL.
R WIRELESS OPERATOR.
F WIRELESS POWERHOUSE
29 Q.M. STOREHOUSE
106 SLEEPING QRS.
201 SUPT. OFFICE (OLD P.O.)
202 ARTES. WELL & ELEC. PUMP.
30 RADIO COMPASS STATION
81 QUARTERS RADIO.
BATTERIES
1. HABERSHAM 4-12" M.
2. FENWICK 1-12" N. Dis.
3. GARLAND 1-12" "
* 4. BRUMBY
5. BACKUS 1-4.7 P.
*6 GANTT
* Dismantled
Wall Broken Down
Approx.
Sea Wall
H.W.
Wall broken down
Wireless Aerial
Walks
Roads
R.R. Sta.
High Water
C. & G. S. Sta.
GARLAND
BRUMBY
FENWICK
BACKUS
GANTT
HABERSHAM
Lt. House Res.
for U.S. Engrs.
Resvn.
Pvt. Whf.
M.B.H.
500 0 1000 2000

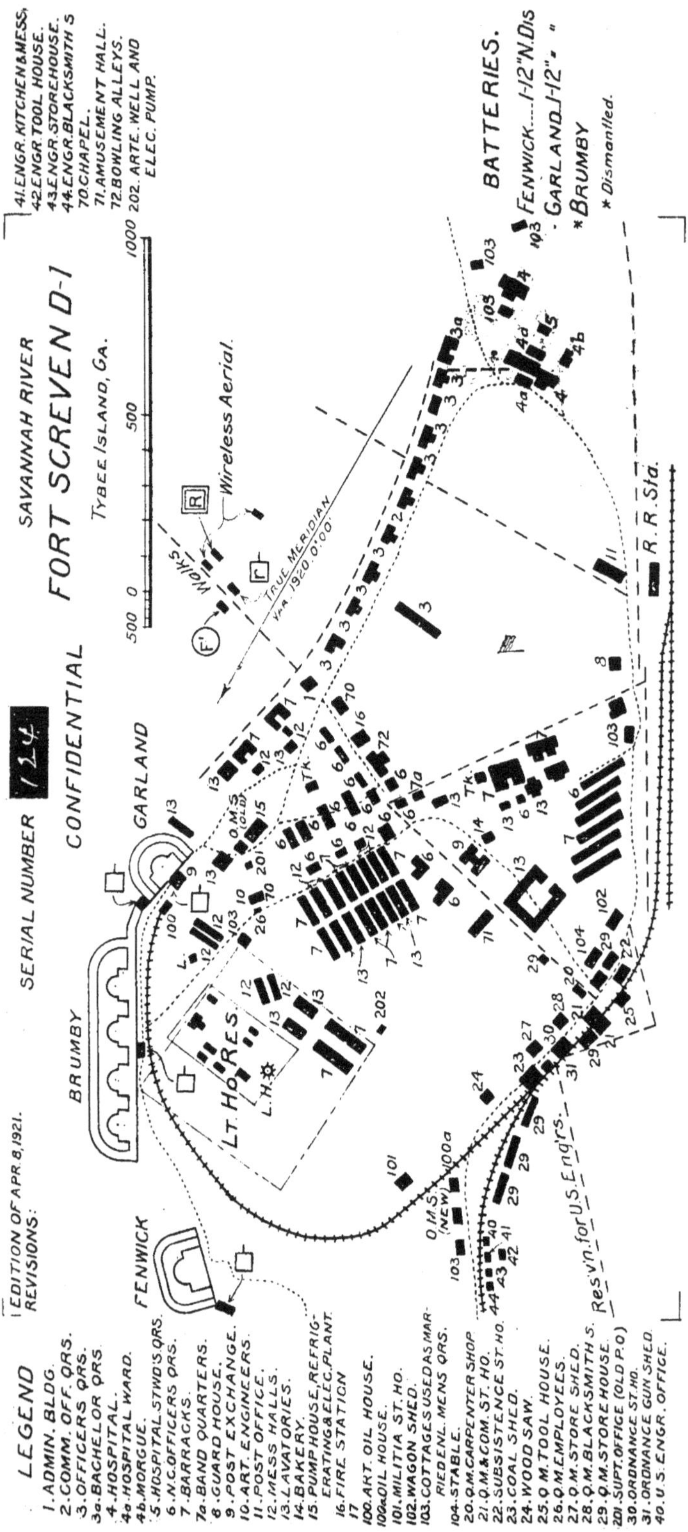
SERIAL NUMBER 124
CONFIDENTIAL
SAVANNAH RIVER
FORT SCREVEN D-1
TYBEE ISLAND, GA.
500 0 500 1000
EDITION OF APR. 8, 1921.
REVISIONS:
BRUMBY
GARLAND
FENWICK
Wireless Aerial.
TRUE MERIDIAN
YAR. 1920 0°00'
LT. HO. RES.
L.H.
R.R. Sta.
Res'vn. for U.S. Engrs.
O.M.S. (OLD)
O.M.S. (NEW)
LEGEND
1. ADMIN. BLDG.
2. COMM. OFF. QRS.
3. OFFICERS QRS.
3a. BACHELOR QRS.
4. HOSPITAL.
4a. HOSPITAL WARD.
4b. MORGUE.
5. HOSPITAL STWD'S. QRS.
6. N.C. OFFICERS QRS.
7. BARRACKS.
7a. BAND QUARTERS.
8. GUARD HOUSE.
9. POST EXCHANGE.
10. ART. ENGINEERS.
11. POST OFFICE.
12. MESS HALLS.
13. LAVATORIES.
14. BAKERY.
15. PUMP HOUSE, REFRIGERATING & ELEC. PLANT.
16. FIRE STATION
17
100. ART. OIL HOUSE.
100a. OIL HOUSE.
101. MILITIA ST. HO.
102. WAGON SHED.
103. COTTAGES USED AS MARRIED ENL. MENS QRS.
104. STABLE.
20. Q.M. CARPENTER SHOP
21. Q.M. & COM. ST. HO.
22. SUBSISTENCE ST. HO.
23. COAL SHED.
24. WOOD SAW.
25. Q.M. TOOL HOUSE.
26. Q.M. EMPLOYEES.
27. Q.M. STORE SHED.
28. Q.M. BLACKSMITH S.
29. Q.M. STORE HOUSE.
201. SUPT. OFFICE (OLD P.O.)
30. ORDNANCE ST. HO.
31. ORDNANCE GUN SHED.
40. U.S. ENGR. OFFICE.
41. ENGR. KITCHEN & MESS.
42. ENGR. TOOL HOUSE.
43. ENGR. STOREHOUSE.
44. ENGR. BLACKSMITH S.
70. CHAPEL.
71. AMUSEMENT HALL.
72. BOWLING ALLEYS.
202. ARTE. WELL AND ELEC. PUMP.
BATTERIES.
FENWICK ..... 1-12" N.Dis
GARLAND 1-12" "
*BRUMBY
* Dismantled.

River. In all, 23 guns and mortars would be mounted in the batteries. The southernmost battery was on the northern tip of Wassaw Island. The battery mounted two 4.7-inch rapid-fire rifles.

During the Spanish-American War, signals were sent from the top of Tybee's lighthouse to the Wassaw Island battery. In 1900, the two guns were removed and remounted at Battery Backus on Tybee Island. Because of erosion, the remains of the Wassaw Island battery are now surrounded by water.

On Tybee Island, the southernmost emplacement is Battery Garland. Constructed near the Tybee lighthouse, Battery Garland contained one 12-inch rifle on a barbette carriage. Construction began in 1898, and the gun was mounted in February, 1899. The gun remained at the battery until 1942, when it was scrapped. The battery was named for a veteran of the War of 1812 and the Mexican War, General John Garland. In 1846 Garland valiantly attacked the Monterrey defenses as his command came under a murderous crossfire. Battery Garland currently serves as the Tybee Museum.

Adjacent to Battery Garland is Battery Brumby which once mounted four 8-inch rifles on disappearing carriages. It was named for Lieutenant Thomas M. Brumby who served at the battle of Manila Bay and died in 1900 from a disease contracted in the Philippines. Begun in April, 1897, and armed by May of I898, this is the only bat-

A view of the main garrision area NARA

tery on the post that was finished in time to serve during the Spanish-American War. The battery required a crew of four officers and 157 men. Today, part of the battery serves as the Tybee Light Shrine Club; another part contains private residences, and Emplacement Number Three is the only intact emplacement. The Tybee Historical Society has made plans to restore this historical emplacement.

Slightly north of Battery Brumby is Battery Fenwick, which once mounted one 12-inch rifle on a barbette carriage. Two officers and 47 men were required to operate the battery, which was named for Colonel John R. Fenwick, a veteran of the War of

Tybee Lighthouse Mark Berhow Collection

Post guardhouse Mark Berhow Collection

Battery Brumby, emplacement #3 Mark Berhow Collection

Side view of Battery Backus Dale Manuel Collection

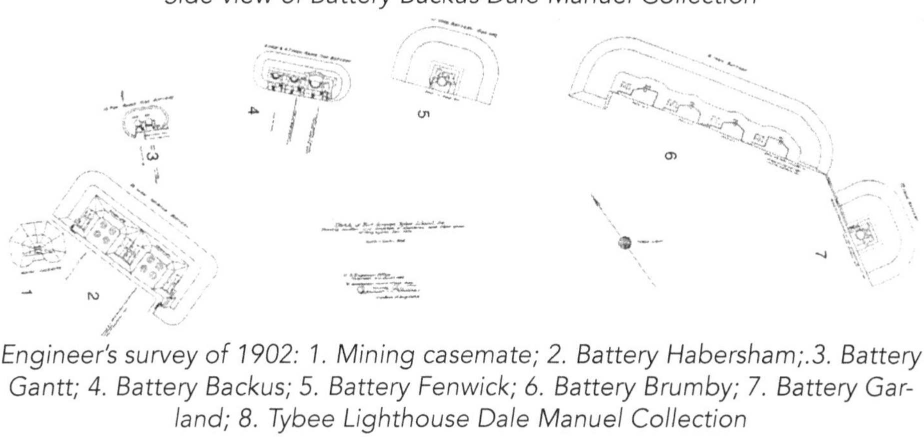

Engineer's survey of 1902: 1. Mining casemate; 2. Battery Habersham;.3. Battery Gantt; 4. Battery Backus; 5. Battery Fenwick; 6. Battery Brumby; 7. Battery Garland; 8. Tybee Lighthouse Dale Manuel Collection

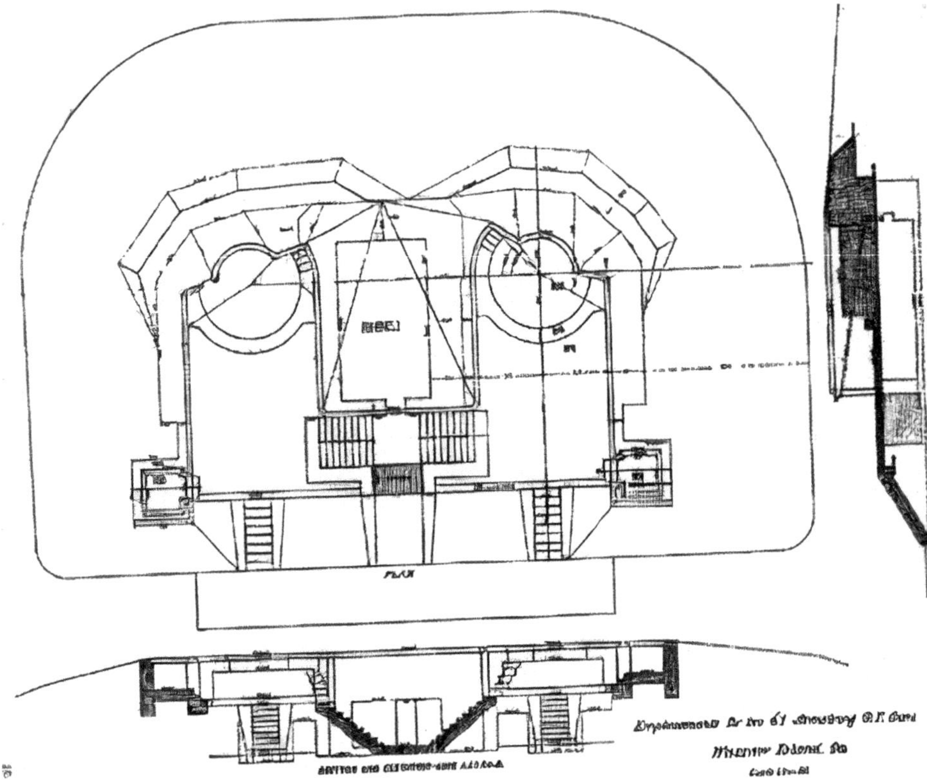

Plan of the battery on Wassaw Island that mounted two 4.7 rifles
Dale Manuel Collection

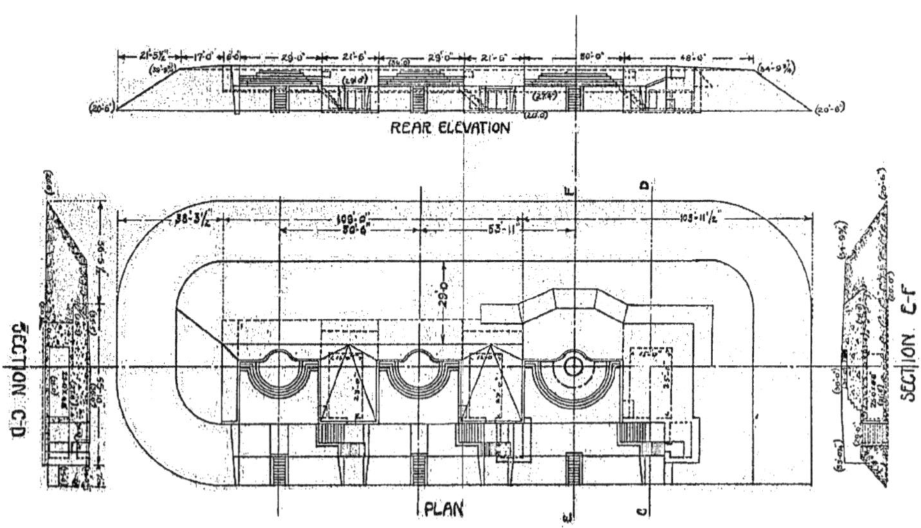

This gun emplacement at Battery Backus once held a 6-inch rifle on a pedestal mount. The 6-inch rifle was later replaced by a 4.7-inch rifle
Mark Berhow Collection, Dale Manuel Collection

1812. During WW II, the gun was scrapped. Today the battery is obscured by a private house built on top of it.

To the north of Battery Fenwick is Battery Backus which originally mounted a single 6-inch Armstrong rapid-fire rifle mounted on September 6, 1898. In 1900, two 4.7-inch rapid fire guns on pedestal mounts were removed from the Wassaw Island battery and were installed on emplacements added to Battery Backus. The 6-inch Armstrong rifle was replaced by another 4.7-inch rifle in 1905. The battery was named for Colonel Electus Backus, who was mortally wounded during the Battle of Sacketts Harbor, New York, during the War of 1812. Today, the battery is on private land with a house built top of it.

Rear view of Battery Backus Mark Berhow collection

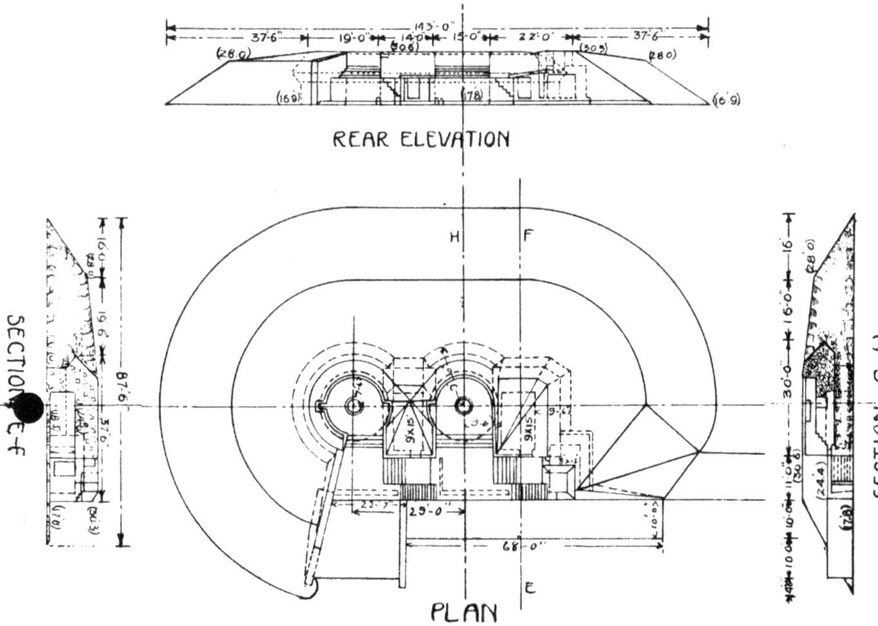

Battery Gantt NARA

Rear view of Battery Gantt Mark Berhow Collection

To the north of Battery Backus is Battery Gantt which once mounted two 3-inch rapid-fire rifles. The battery was named for First Lieutenant Levi Gantt, who was killed in the storming of Chapultepec Castle outside of Mexico City during the Mexican War in 1847. The battery was completed in February of 1900 and armed in September of 1903. The battery required a crew of one officer and twenty-five men. Today, the battery is intact and on private land, with a house built on top of it.

Part of the massive Battery Brumby is now the Shrine Club
Mark Berhow Collection

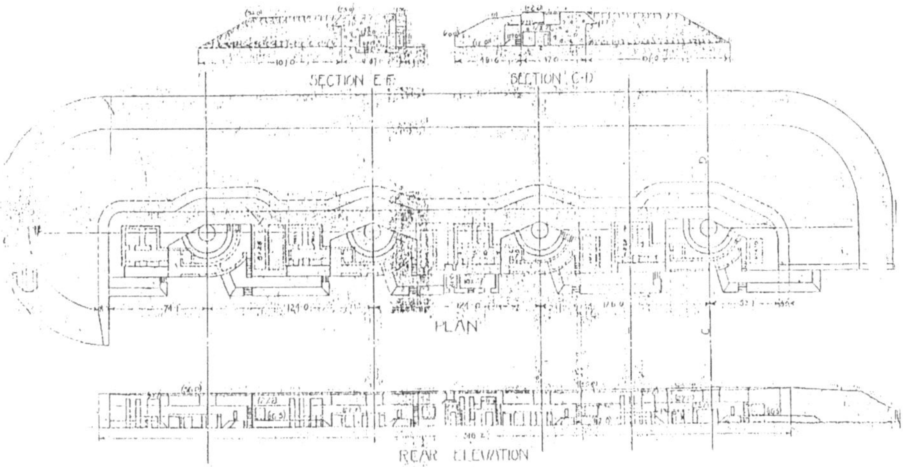

Battery Brumby NARA

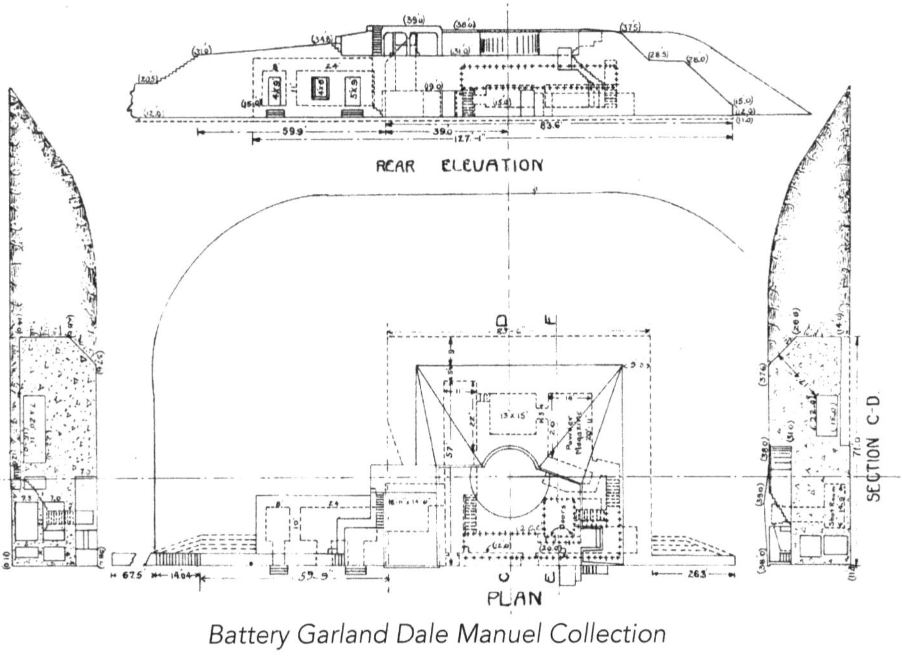

Battery Garland Dale Manuel Collection

Behind Battery Gantt is Battery Habersham with two mortar pits, each armed with four 12-inch mortars. Construction started in December of 1898 and the mortars were mounted by June of 1900. A crew of 7 officers and 219 men was required to operate the battery. Its namesake, Major Joseph Habersham, served during the Revolutionary War and was Postmaster General of the United States. During WW I, the mortars were removed and sent to France for field artillery. After the war, the two pits were used for coal bins. Today, one of the pits is grown over with trees and weeds, while the other is used for storage by the local cable company.

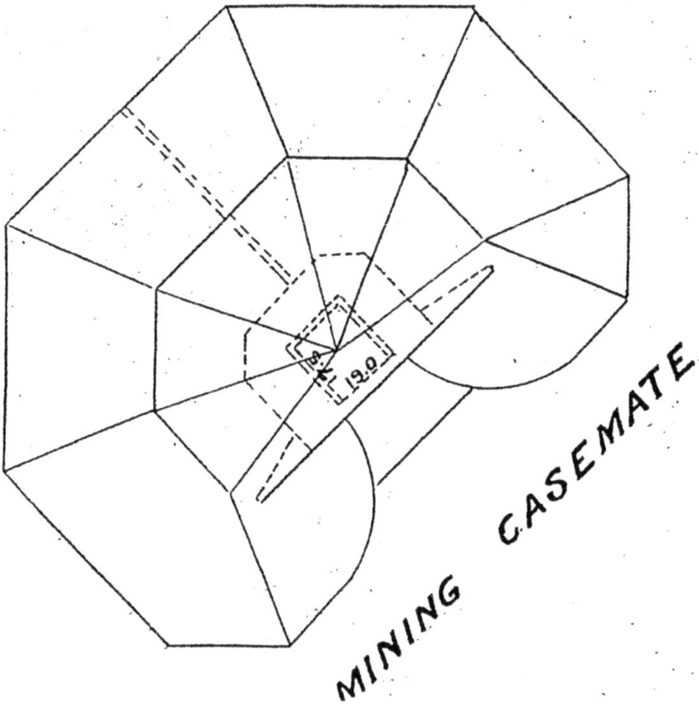

Mining Casemate Dale Manuel Collection

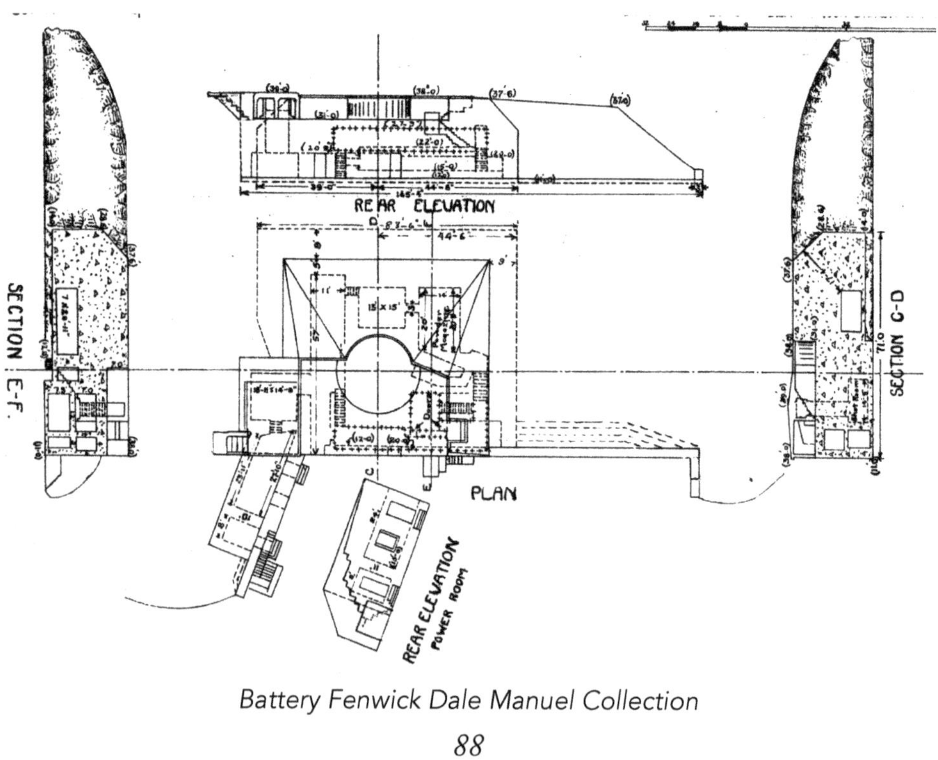

Battery Fenwick Dale Manuel Collection

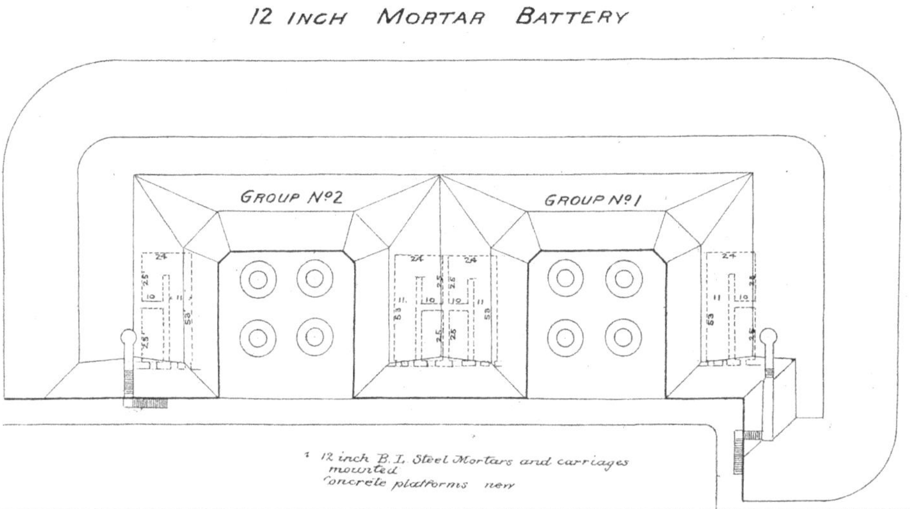

Battery Habersham a) drawing and b) photo
Dale Manuel Collection, Mark Berhow Collection

On the north shore of Cockspur Island, near Fort Pulaski, is Battery Hambright, which once contained two 3-inch rapid-fire rifles that guarded the north channel minefield. Construction began in June, 1899. The battery is now intact and administered by the National Park Service, along with nearby Fort Pulaski.

In 1898, Fort Pulaski became the mine depot of the north channel. A mine casemate was constructed inside the demilune. Under casemates four and five of the old fort, cable tanks were constructed to hold sixteen cable reels.

A mine depot was also located on Tybee Island. There was once a mine casemate to the rear of Battery Habersham and nearby were torpedo-storage and dynamite-storage buildings. The mine wharf was located on the Tybee Roads side of the reservation.

During the "Great Depression," Fort Screven served as headquarters of the District F, IV Corps of the Civilian Conservation Corps. Also during the 1930s, the fort was used as a training site for the Civilian Military Training Corps. In 1932, the garrison numbered about 400 and the post commander was Lieutenant Colonel George Catlett Marshall, who is best known as Chief of Staff during WW II and author of the "Marshall Plan," a plan for the rebuilding of war-torn Europe after WW II.

The post grew in size over the years until it was sold in 1945. The buildings and facilities consisted of officers' quarters, barracks, two parade grounds, a rifle range, a guard house, a post theater, a bakery, a bowling alley, shops, garages, a barber shop, a gymnasium, a fire station, chapel, commissary, P/X, restaurant, laundry, veterinarian office, stables, library, blacksmith shop, a sewage system, a morgue, an infirmary, a water system, storage buildings, a post office, a bandstand, a telephone exchange, storage tanks, and a recreation building.

During WW II, Fort Screven became a diving school for underwater salvage and port repair. In October, 1944, Fort Screven was declared surplus. On November 27, 1945, it was sold to the town of Savannah Beach. The old fort area became a mix of privately and city owned property. Many historic buildings still remain.

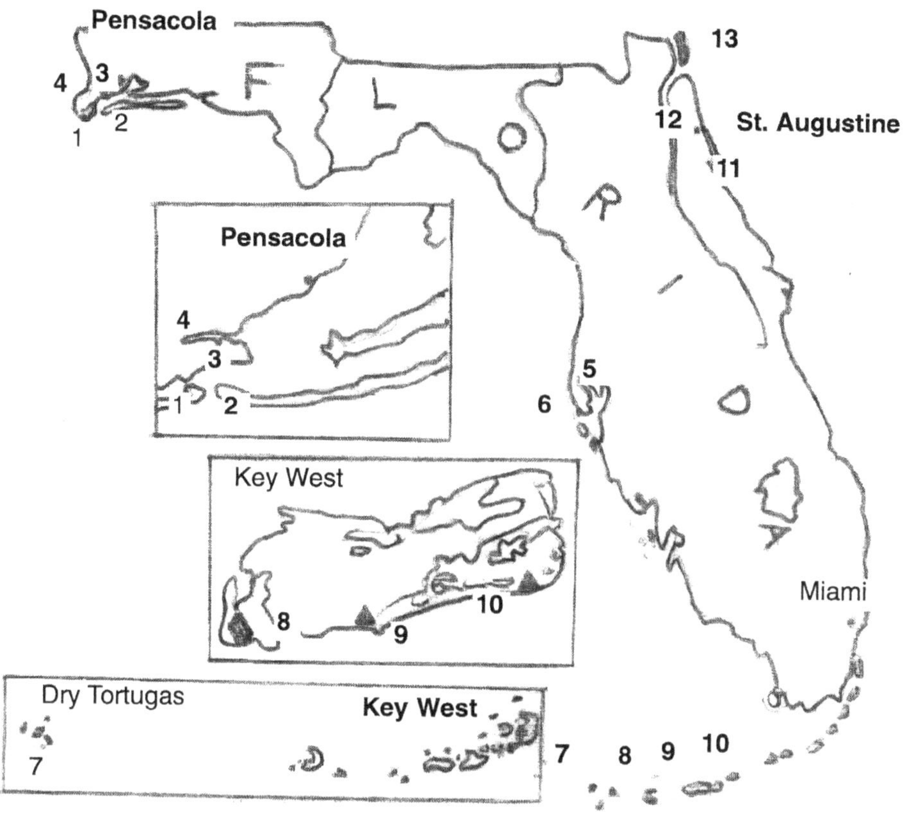

269. Forts of Florida 1. Fort McRee; 2. Fort Pickens; 3. Fort Barrancas; 4. Advanced Redoubt; 5. Fort De Soto; 6. Fort Dade; 7. Fort Jefferson; 8. Fort Taylor; 9. West Martello Tower; 10. East Martello Tower; 11. Fort Matanzas; 12. Castillo de San Marcos; 13. Fort Clinch Dale Manuel Collection

Coastal Defenses of Pensacola Bay Florida

Fort McRee, Foster's Bank, Pensacola Bay Begun 1833

Fort McRee was the second of four forts eventually built to defend Pensacola Bay and its Navy Yard at Tartar's Point. Begun in 1833 and finished in 1838, it was constructed on Foster's Bank, a narrow sand bank across from Fort Pickens on Santa Rosa Island. It was built along with an external battery to command the main ship channel at the entrance to the bay. The fort was named for Colonel William McRee of the Army Corps of Engineers and an original member of the Fortifications Board. Colonel McRee resigned from the Army in 1819 because he resented the appointment of the French engineer, General Simon Bernard. Ironically, it would be General Bernard who would design the fort that would be named in McRee's honor.

The fort was an elliptical castle designed to fit on the narrow bank. Castle Calhoun (Ft. Wool) in Hampton Roads, Virginia, is similar in plan. Fort McRee had three tiers of guns, with 23 barrel-vaulted casemates, and a barbette tier. Under each

Sketch of Fort McRee Dale Manuel Collection

massive arch were two tiers. The tiers were separated by a wooden platform of planks and timbers. On the sea faces and curved ends, each tier had two guns per arched casemate. In the gorge each tier held one twenty-four-pounder flank carronade.

The fort was designed for 108 guns and 350 men in time of war. The channel casemates could hold 44 guns, and the barbette tier that faced the channel could hold 25 guns. The armament for the land side could mount as many as 39 guns.

The main work of the fort was 35 feet high, 146 feet wide and 450 feet long. There was a total of twenty-three casemates. Eleven casemates were in the two sea faces. The casemates on the gorge were smaller and were enclosed for quarters. Two magazines were located at each end of the gorge. Two cisterns on each side of the gorge provided water. Each circular end contained a magazine and two staircases. At

Plan of Fort McRee. Legend: 1. Magazine; 2. Store Magazine; 3. Hot Shot Furnaces; 4. Quarters; 5. Stair Tower; 6. Cisterns; 7. Sallyport Dale Manuel Collection

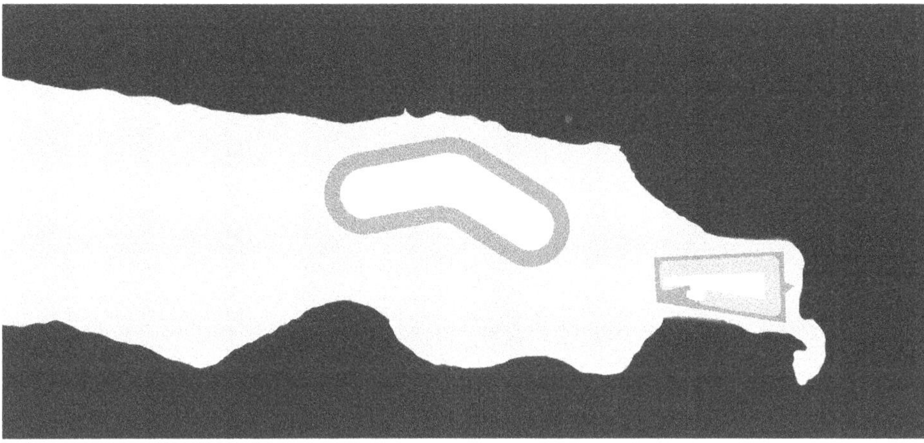

Shown is the layout and position of the South Water Battery in relation to Fort McRee. The battery was designed and positioned in such a way that the guns of Fort McRee could add flanking fire to its sides. John Weaver Collection

the salient angle of the two sea faces was a stair tower with a spiral staircase. Iron stairs were at each end of the gorge.

The two sea faces, as well as the gorge, formed angles of 130 degrees. Casemate guns in the gorge fired downward and could create a flanking crossfire at this angle, thus protecting the sallyport in the middle of the gorge. Above the sallyport was a machicoulis from which projectiles could be dropped or muskets fired down.

Located 440 feet to the south of Fort McRee was the South Water Battery. It was a trapezoid measuring 120 feet wide and 278 feet long. It was an open battery with a designed armament of thirteen guns en barbette. In one corner was a sallyport and a single magazine. Built to bring more firepower to bear on the channel, it had musket loopholes that ran the perimeter of the work. The work was a Carnot Wall design with its detached scarp. Another battery to the north of Fort McRee was planned but never built.

The South Water Battery had only one flanking feature, a small bastionette in the middle of the southern wall. The battery was ingeniously designed and positioned so the guns of Fort McRee could sweep its sides with flanking fire.

Fort McRee, like most of the seacoast forts, was lightly manned for years. When Florida and Alabama troops seized the fort in January, 1861, only a caretaker was there. At that time only 125 cannon were on hand, and most of those not mounted.

Union troops were still in possession of Fort Pickens, across the channel on Santa Rosa Island. Confederates held Fort McRee, Fort Barrancas, and the Advanced Redoubt on the mainland. Fort McRee and Fort Barrancas were poorly located for an attack on Fort Pickens, as both had only been designed for channel defense. Fort McRee could only bring two guns to bear on Fort Pickens. Fort Pickens, however, had command of Fort McRee. Fort Pickens' entire channel-bearing fronts could be used to bombard Fort McRee.

On October 9th, 1861, the battle of Santa Rosa Island ended in the retreat of the Confederates back to the mainland. In retaliation for the attack, on November 22, soldiers at Fort Pickens as well as men on board two ships, the Richmond and the Niagara, opened fire on Fort McRee, the Navy Yard, and the other main land batteries. After two days, 5,000 rounds had been fired by Union troops and 1,000 by Confederate troops. Fort McRee, Fort Barrancas, and the Navy Yard were severely damaged.

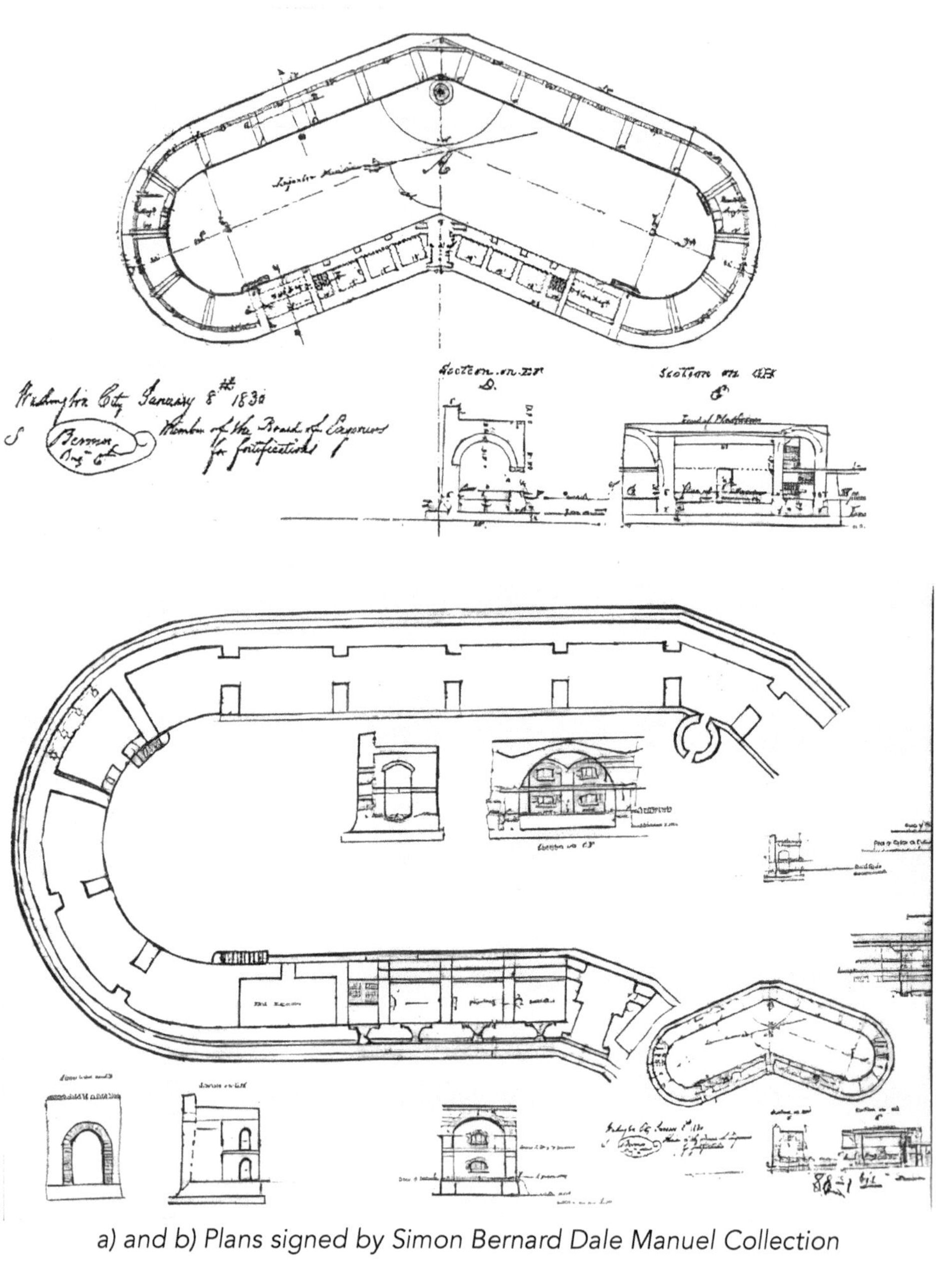

a) and b) Plans signed by Simon Bernard Dale Manuel Collection

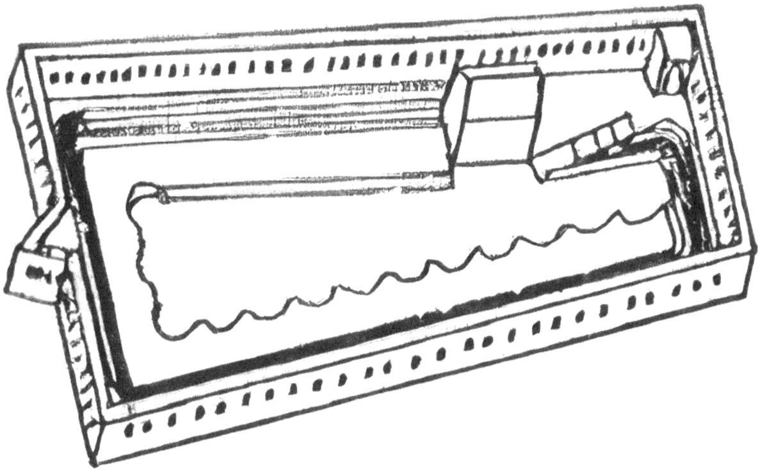

Drawing of the South Water Battery Dale Manuel Collection

Drawing of Fort McRee Dale Manuel Collection

Volleys riddled the scarp of Fort McRee and set many fires that threatened the magazines. After the bombardment, it was suggested that Fort McRee be abandoned. Confederate commander General Braxton Bragg decided it would be bad for morale and ordered the fort to be held.

Another exchange of fire occurred on January 1, 1862. Pressing needs in Tennessee caused the withdrawal and evacuation of Pensacola on May 9th, 1862. The evacuating Confederates filled the casemates with everything that was combustible at Fort McRee and set the fort on fire. The infrastructure of the fort was destroyed, leaving nothing but the bare walls. The Union troops occupied the bay defenses for the rest of the war.

Interior of Fort McRee during the Civil War Dale Manual Collection

Fort McRee and the South Water Battery are shown in this drawing. Harper's Weekly

Postcard showing the ruins of Fort McRee Dale Manual Collection

During its existence, Fort McRee fought a losing battle with the sea. During the 1840s and 1850s, jetties were built to keep the sea from encroaching on the fort. Foster's Island and Bank have migrated both east and west many times over the years. Now an independent island does not exist; is now part of Perdido Key.

Storms and hurricanes have left the fort in ruins. Now even the ruins have vanished. Two Endicott batteries named Battery Slemmer (two 8-inch guns) and Battery Center (four 3-inch rapid-fire guns) were built nearby. A seawall was constructed around the concrete installations. In WW II, the two 6-inch shield gun Battery # 233 was constructed between the two Endicott batteries but was not armed.

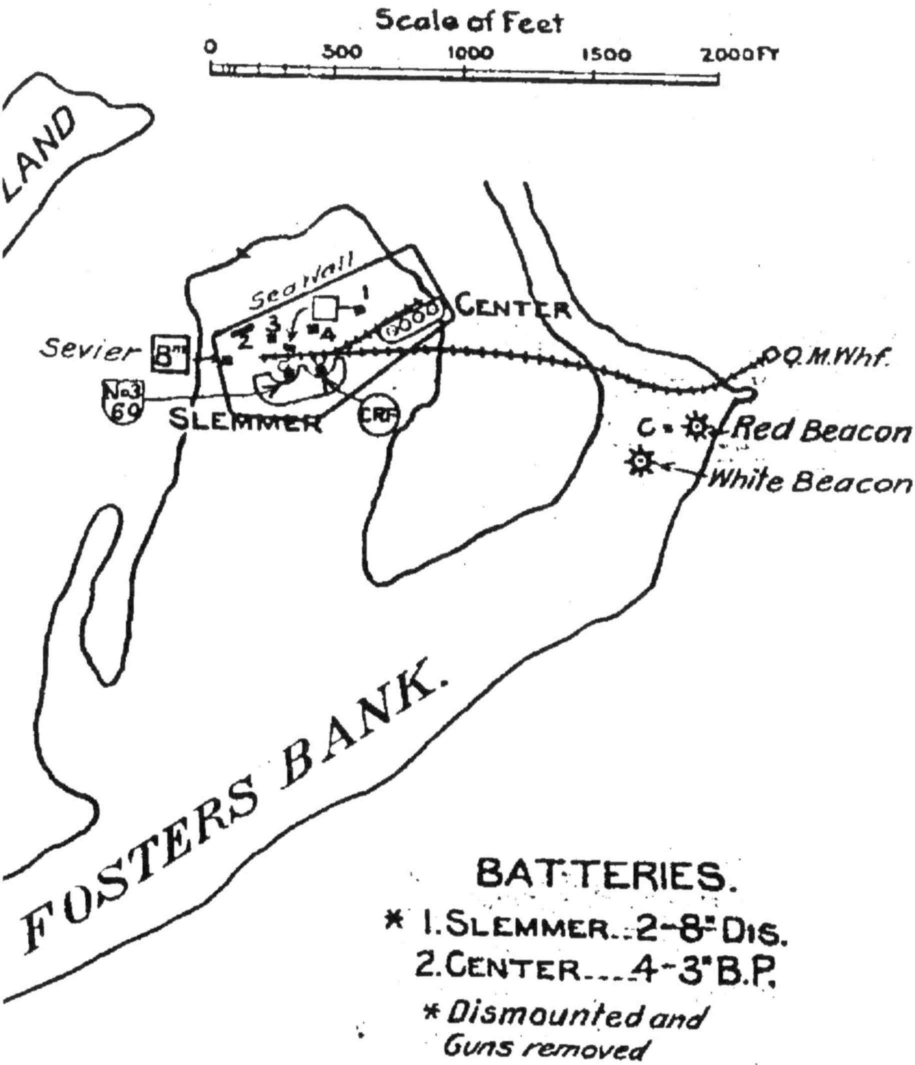

Site plan of the Endicott batteries at the Fort McRee Reservation. Legend: 1. Quarters; 2. Barracks; 3. Engr. Store House; 4. Q.M. Coal Shed
Dale Manuel Collection

The entrance to WWII Battery #233.

The almost-buried concrete parapet of Battery Slemmer Mark Berhow Collection

The remains of an instrument column and the seawall that surrounds the area Dale Manuel Collection

The remains of Battery Center Mark Berhow Collection

Fort Pickens, Santa Rosa Island, Pensacola Bay Begun 1829

The Fortifications Board chose the western end of Santa Rosa Island to build Fort Pickens, the first in a series of four forts to guard Pensacola Bay and the Navy Yard. Fort Pickens was the largest of these forts and was started in 1829 and finished in 1834.

Sketch of Fort Pickens Dale Manuel Collection

Fort Pickens was built to guard the mouth of the bay. A large pentagon with five bastions, the fort measured 550 feet x 350 feet x 37 feet tall. It has two tiers, with casemated guns on the first tier and guns mounted en barbette on the upper tier. Each casemate had a large, barrel-vaulted arch with embrasures for two guns.

The fort's designed armament was 217 guns. The work included two channel fronts that were about two hundred yards from the water. Today, the island has grown three quarters of a mile beyond the fort. In the middle of the channel fronts was a tower bastion which is slightly taller than the adjacent curtains. Gun embrasures were on each face of the tower bastion and howitzer embrasures were located on each flank. On each end of the channel fronts were bastions that would flank the fronts and

Channel-side curtain and tower bastion. On top of the bastion is a 15-inch Rodman gun Photo by John Weaver

One of two massive gorge bastions that swept the gorge and flank curtains. Both bastions contained mining tunnels and countermine positions. Photo by John Weaver

wrap around to flank the secondary fronts. Two massive bastions were on each end of the gorge, where it intersected the secondary fronts. A dry ditch surrounded the work on three sides, leaving the channel fronts open to the water. On the east, a tall counterscarp with a covered way and glacis bolstered its landward defenses. On each end of the counterscarp was a rifle gallery, added for flanking fire on the sallyport and postern.

In these very large landward gorge bastions were powder magazines. One unusual feature of both gorge bastions was that each contained three mine chambers that could hold 1000 pounds of black powder. These chambers could be exploded if the bastions were taken by the enemy, The defenders could regroup in other areas of the fort. This would be a "last ditch" effort in the defense of the fort.

Large casemates on the channel front contained two gun emplacements each Photo by John Weaver

The sallyport and damaged curtain. The northwest bastion was blown up by an accident in 1899 Photo by John Weaver

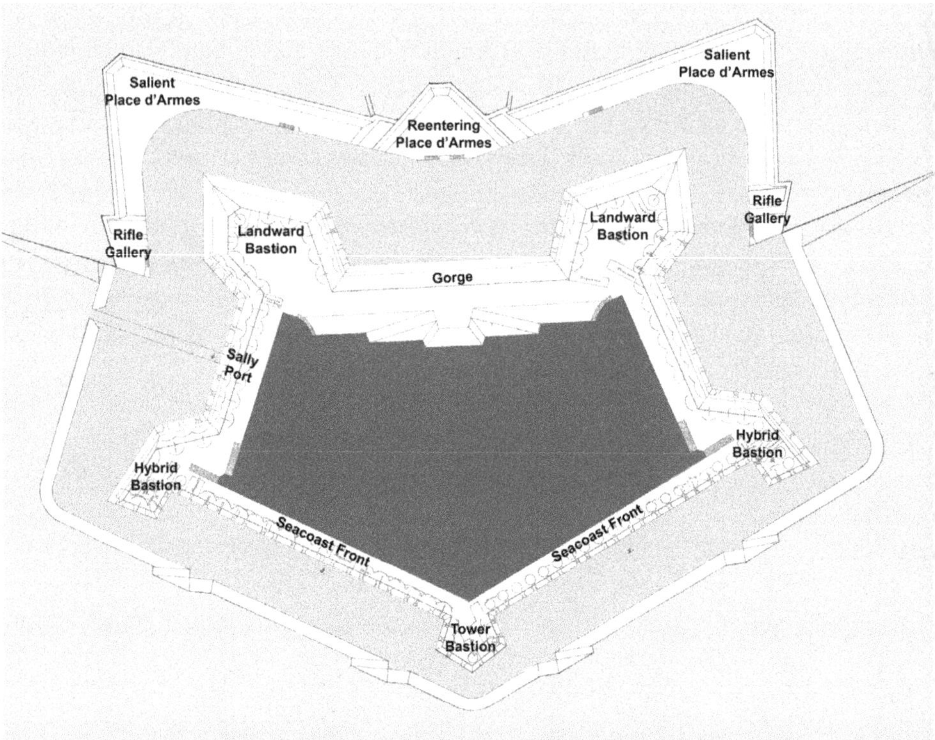

Plan of Fort Pickens and outworks John Weaver Collection

The gorge bastions contained casemated gun rooms and magazines. During the Civil War, blindages were built in front of these magazines to block incoming shot and shells. On the second tier of these bastions were guns mounted en barbette.

Fort Pickens was named in honor of General Andrew Pickens, a Revolutionary War hero. The fort was originally designed by Joseph Totten, who began construction on the seacoast fronts prior to getting the approval of General Bernard. Bernard allowed the design of the seacoast fronts to stand, but greatly modified the landward sides of the fort, reducing its size substantially. The money saved by Bernard's modifications was used to fund the three other Pensacola forts. The final design was vindicated during the action that occurred here during the Civil War. The fort's position and design made the Confederate grip of the mainland forts impractical. William Chase was the on-site Army engineer that supervised construction of Fort Pickens and the other Pensacola Bay defenses.

Fort Pickens was one of a handful of forts in the South that remained in Union hands throughout the Civil War. On the night of January 10, 1861, Lieutenant Adam Slemmer moved his small command of fifty soldiers and thirty sailors from Fort Barrancas to Fort Pickens. Florida and Alabama troops occupied the rest of the installations on the bay. Ironically, the officer that demanded Fort Pickens' surrender to the State of Florida was the fort's builder, William H. Chase. Lieutenant Slemmer refused to surrender and held out until reinforcements landed in April.

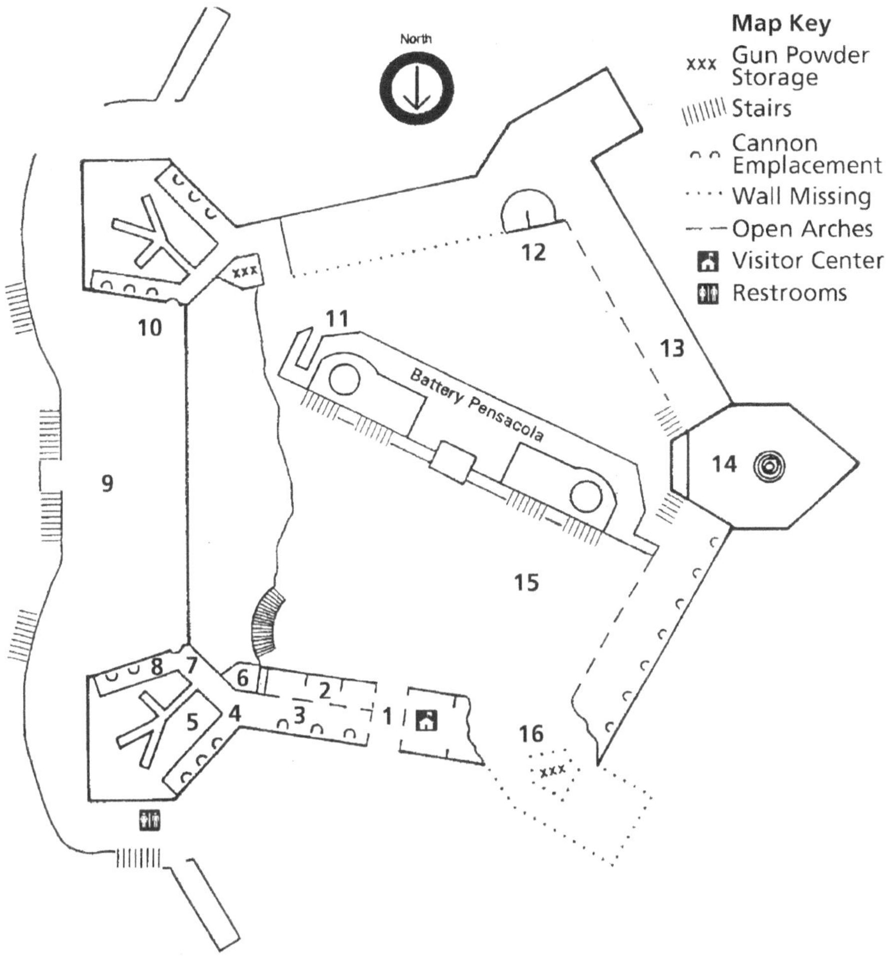

Plan of Fort Pickens National Park Service

By the summer of 1861, the Confederate ranks grew to 8,000 men under General Braxton Bragg; the Federal Troops at Fort Pickens were only about 2,000 men under Colonel Harvey Brown. On September 2, 1861, troops from Fort Pickens burned a floating dry dock in the bay. On September 13, naval forces from Fort Pickens raided the Navy Yard docks and burned the schooner Judah. In retaliation, the Confederates landed a force of 1,500 men on Santa Rosa Island the night of October 9th. The Confederates overran the camp of the 6th New York Volunteer Regiment. When troops from Fort Pickens reinforced the Union ranks, the Confederates retreated to their boats. In the Battle of Santa Rosa Island, the Union had 14 killed, 29 wounded and 24 captured. The Confederate losses were 87 men.

On November 22 and 23, 1861, the garrison of Fort Pickens began a bombardment of the Confederate- held installations. Both Fort McRee and the Navy Yard were heavily damaged. Another bombardment occurred on January 1, 1862, and the Navy Yard was further damaged. With Fort Pickens in a commanding position on Santa Rosa Island, the Navy Yard and Fort McRee in ruins, and with more Confederate

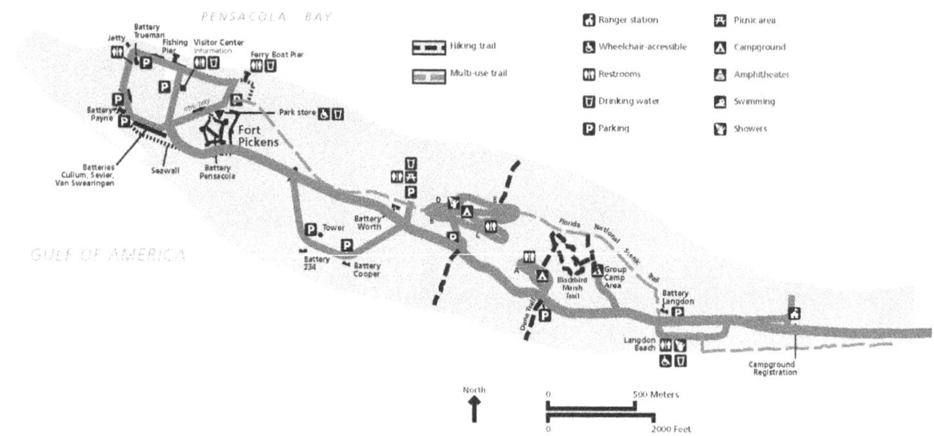

The location of various batteries on the Fort Pickens Reservation National Park Service

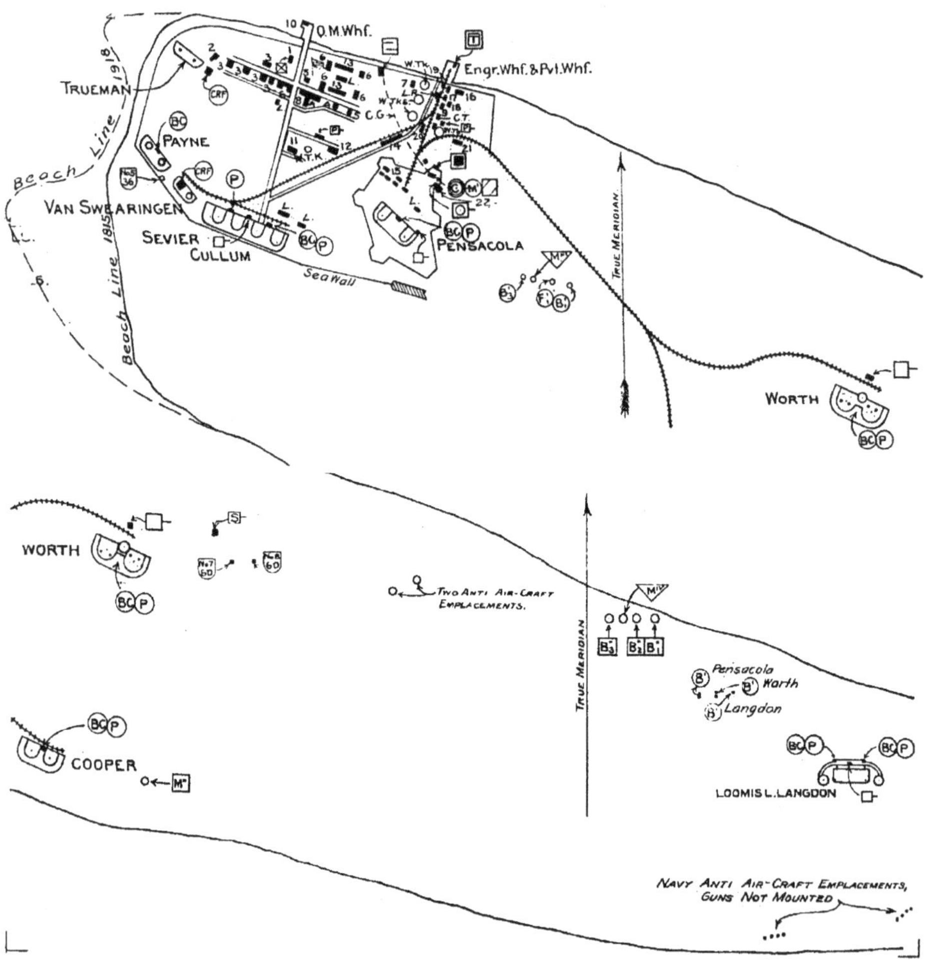

Fort Pickens Reservation 1921 NARA

Battery Van Swearingen once mounted two 4.7-inch rapid-fire rifles
Mark Berhow Collection

Buildings that remain on the Fort Pickens Reservation Photo by John Weaver

Harbor Defense Command Post on top of Battery Worth
Mark Berhow Collection

In 1937, 155mm rifles like this one were mounted on Panama mounts near Battery Cooper Mark Berhow Collection

Dual range finding and Battery Commanders station for WWII Battery #234 The concrete top has since been removed and placed on the ground next to the steel tower Photo by John Weaver

Batteries Cullum and Sevier once held four 10-inch "disappearing" rifles. In WWII, Battery Trueman's 3-inch rifles were removed and mounted here
Mark Berhow Collection

One of the 3-inch rapid-fire rifle emplacements of Battery Trueman
Mark Berhow Collection

Battery Langdon, built in 1917 as an open battery for two 12-inch rifles, was casemated in WWII Mark Berhow Collection

Battery Pensacola was constructed inside the old fort's parade. It once mounted two 12-inch rifles on "disappearing carriages." Mark Berhow Collection

This 6-inch shield gun is one of two mounted in Battery #234. These guns are just for demonstration purposes because Battery #234 was never armed. Photo by John Weaver

troops needed in Tennessee, a withdrawal was ordered in May, 1862. With the evacuation of the Confederates, Union troops gained control of the bay. For the rest of the war Fort Pickens served as a prison. The Navy Yard served as the headquarters for the blockading squadrons in this part of the Gulf of Mexico.

After the war, the fort was in caretaker status. During the 1870s it was used as a military prison. From October, 1886 to May, 1888, Geronimo and many Chiricahua Apaches were housed in the fort.

During the 1890s, a large Endicott battery, named Battery Pensacola, was built on the fort's parade. It mounted two 12-inch rifles on disappearing carriages. The Army lowered the walls of the southwest sea face and bastion of old Fort Pickens because it obstructed the field of fire of the new battery. In so doing, damage was caused to the southern curtain. The fort was activated for the Spanish-American War in 1898. In 1899, an out-of-control fire caused an explosion in the magazine of the northwest bastion. The bastion was completely destroyed.

Other batteries were built between 1898 and 1906 on Santa Rosa Island. Batteries Cullum and Sevier mounted four 10-inch rifles on disappearing carriages. Battery Van Swearingen held two 4.7-inch rapid-fire rifles. Batteries Payne and Trueman each contained two 3-inch rapid-fire rifles. Battery Worth mounted eight 12-inch mortars. Battery Cooper mounted two 6-inch rifles on disappearing carriages. Most of the frame buildings on the post were constructed during the Endicott period.

The fort was the center for the mining of the harbor entrance. Started in 1917 and completed in 1923, Battery Loomis Langdon mounted two 12-inch rifles en barbette. In WW II, the battery was casemated. In 1937, four Panama mounts with 155mm rifles were constructed around Battery Cooper. Battery #234 was completed in 1943, but never armed. During WW II, the two 3-inch guns of Battery Trueman were moved to Battery Cullum. The Harbor Defense Command Post was constructed on top of Battery Worth. Two fixed 90mm batteries were mounted on the other side of the seawall near the old fort. Four mobile 3-inch anti-aircraft batteries were erected between the old fort and Battery Cullum. Two fixed 3-inch anti-aircraft batteries were constructed between Batteries Worth and Langdon. A signal station and a signal mast with an elevation of 85 feet was erected on top of Battery Sevier.

The fort was active in both World Wars as an artillery and anti-aircraft post. Fort Pickens and the Fort McRee Reservations were occupied by the 13th Coastal Artillery, headquartered at Fort Barrancas. In the 1920s and '30s it was also a training camp for the Florida National Guard, ROTC, and Citizens Military Training Corps. The Civilian Conservation Corps used the fort as a regional headquarters. Fort Pickens became a part of Gulf Islands National Seashore, a unit of the National Park Service.

Fort Barrancas, Pensacola Bay Begun 1840

Built on the site of the old Spanish Fort San Carlos de Barrancas, Fort Barrancas is situated on a high bluff, or barrancas, overlooking Pensacola Bay. Construction began in 1840 and was completed in 1844. Below the bluff was the old Spanish Bateria San Antonio, built in 1797. This battery was refurbished in 1840 to serve as a water bat-

Photograph of Fort Barrancas gun line John Weaver Collection

tery for Fort Barrancas. Fort Barrancas is a large quadrilateral made from six million bricks. The main work's curtains measure 182 feet x 195 feet x 120 feet x 125 feet. Fort Barrancas was designed by Joseph Totten. Fort Livingston, Louisiana, is very similar in shape. The on-site engineer at Fort Barrancas was William Chase, who was also the builder of Fort Pickens and Fort McRee.

Fort Barrancas was the third of the Pensacola Bay defenses to be built. It was built to work in conjunction with Fort Pickens across the bay and Fort McRee located diagonally at the mouth of the bay. The crossfire from these forts would have been devastating. When constructed, Fort Barrancas was directly aligned with the bay entrance, but is no longer now due to the westward migration of Santa Rosa Island. Fort Barrancas was also the southern anchor of land-side defensive entrenchments that ran north 700 yards to the Advanced Redoubt.

A single sallyport gives entrance to the work over a drawbridge. Guard rooms, storerooms, and magazines are on both sides of the sallyport. Off the sallyport are entrances to the musket galleries that encircle the entire work. Also off the sallyport is a subterranean tunnel that leads to the counterscarp gallery. The sallyport opens to the parade.

Only a hot-shot furnace, well, and ramps that ran to the terreplein were on the parade. At the fort's parapets were guns mounted en barbette. The fort's parapets rise to an elevation of 69 feet above sea level. Guns were mounted on the seacoast fronts and land fronts. At the salient where the seacoast fronts meet, a subterranean tunnel leads to the old Spanish water battery.

The old Spanish water battery is a semicircular work. Built of brick and stucco, it was designed to mount thirteen guns en barbette. The water battery, being at a lower level than the main work of Fort Barrancas, could sweep the bay, skipping shot to the water line of ships. At the gorge of the battery is a reduit or bombproof that contains storerooms and a powder magazine. The magazine had a wall or "blindage" around its entrance to block incoming shot and shell. A stairway led to a rifle platform at the top of the work.

The main work and the counterscarp Photo by John Weaver

In front of the main work is the Spanish water battery Photo by John Weaver

Fort Barrancas and water battery from the bottom of the bluff
Photo by John Weaver

Corner where the two counterscarp walls meet Photo by John Weaver

The musket galleries that encircle the main work have loopholes with vent holes above them. The vent holes were to ventilate the black power smoke of fired muskets. The galleries have vaulted arches and retaining walls to buttress the weight of the above armament and earthen fill.

A dry ditch is on the two land fronts of the work. Lining the dry ditch on these two fronts is the counterscarp with vaulted musket galleries that ran its length. A total of eight counterfire rooms are in the counterscarp with four ready magazines. Two flank howitzers were mounted on each end and on the reentering angle where the two galleries intersect. These guns could sweep the two land fronts of the dry ditch and the two seacoast fronts of the main work. Two posterns gave access to the ditch from the counterscarp gallery.

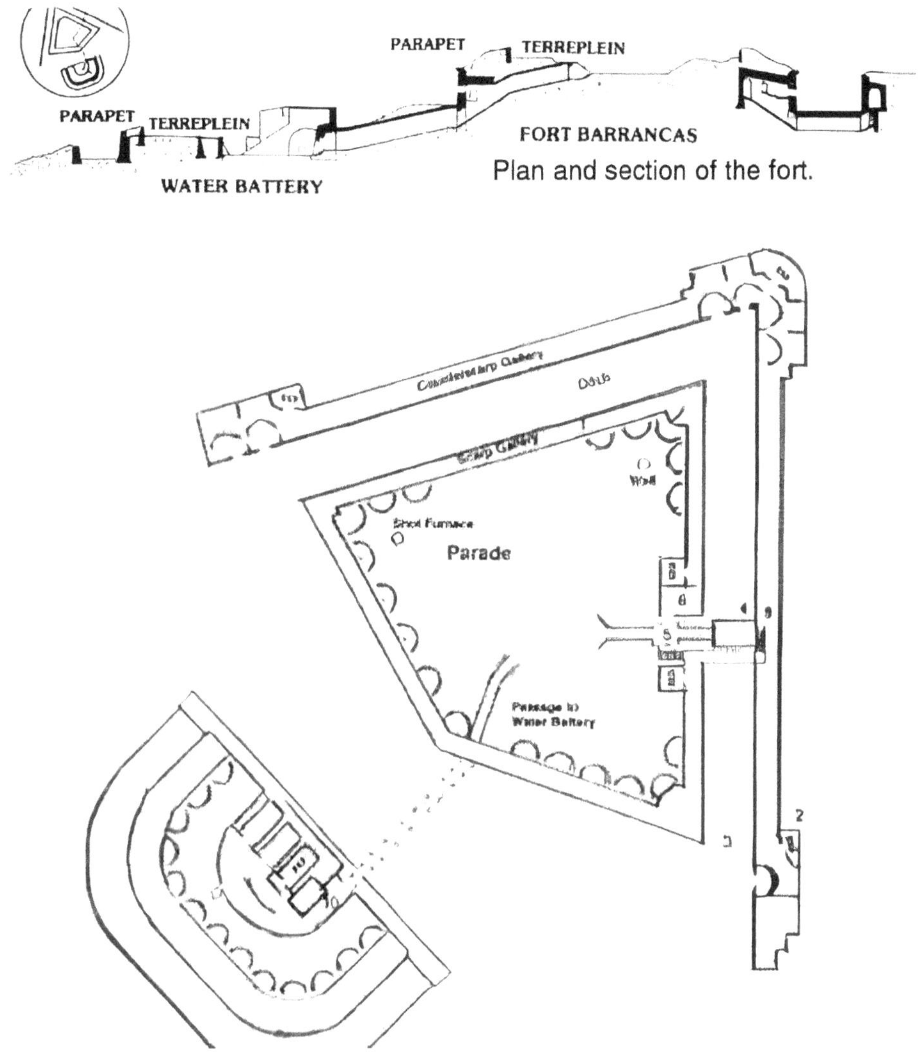

a) and b) Fort Barrancas 1840 Plan and section of the fort. National Park Service.

On the eve of the Civil War, only 50 Union soldiers were at Fort Barrancas. When Confederate skirmishers approached the fort to take command, shots rang out. These were the first shots of the Civil War. While the skirmishers were driven off, the small garrison could not hold the mainland forts in the long term, so Lt. Adam Slemmer moved his command to Fort Pickens on Santa Rosa Island. After the Union evacuation, Fort Barrancas, Fort McRee, the Advanced Redoubt, and the Navy Yard were controlled by Confederates.

The Confederate garrison of Fort Barrancas participated in bombardments of Fort Pickens and batteries on November 22 and 23, 1861, and on January 1, 1862. By May 1862, the Confederates withdrew from Pensacola Bay and left the military installations to the Union Army for the duration of the War.

The Reduit of the old Spanish water battery Photo by John Weaver.

In the 1840s, barracks and officers' quarters were built. In 1847, a new barracks was constructed, called Barrancas Barracks. Three stories high, 36 feet wide and 186 feet long, this brick barracks could house 400 men. During the 1850s and 1870s, quarters and a hospital were added as the post of Fort Barrancas grew.

Fort Barrancas was the headquarters for the 13th Coast Artillery Regiment. The regimental batteries, located on Santa Rosa Island and Perdido Key, were constructed between 1894 and 1942. The fort was a large artillery training center by the 1930s. Army garrisons occupied the post until it was declared surplus in 1947. Most of the buildings of the post are used today by the Pensacola Naval Air Station. The National Park Service restored Fort Barrancas and the water battery in the late 1970s, and is a part of the Gulf Islands National Seashore.

Advanced Redoubt, Pensacola Bay Begun 1845

The Advanced Redoubt was the fourth and last of the Pensacola Bay fortifications. Built 700 yards north of Fort Barrancas to anchor a line of entrenchments between the two forts, its purpose was to protect the land approach to the Navy Yard. The Navy Yard was thought to be vulnerable across the peninsula from Fort Barrancas to Bayou Grande. Construction began in 1845, and work dragged on for fifteen years, until finished in 1859.

The fort is a trapezoid. Its enceinte, or main work, measures 182 feet x 148 feet. It is surrounded by a dry ditch. In the center of the ditch is a cunnette, or brick-lined trench used for drainage. The ditch is surrounded by a counterscarp, on which three sides were lined with musket galleries. Two counterfire rooms are on both ends of the

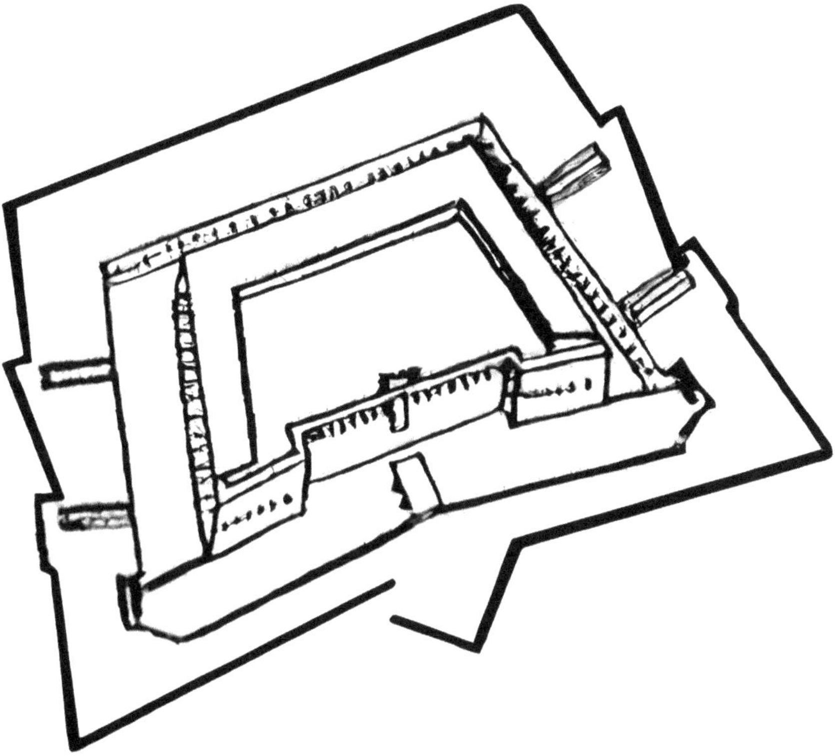

Plan of Advanced Redoubt Dale Manuel Collection

west counterscarp gallery to flank the southern and northern ditches. Other counter-fire rooms are on the northwest corner and commanded the western ditch. A covered way and glacis surround the entire work.

The main work was constructed of brick. The fort's foundation was one of the first to use concrete in its construction. A narrow-gauge railroad transported materials and supplies on tracks that ran 4,000 feet from a wharf on the bay. The first tier had no loopholes, but its vaulting was used to buttress the wall against the weight of the sand soil system.

The Advanced Redoubt from the air. National Park Service

Looking down the west ditch toward the north counterscarp gallery
Photo by John Weaver

The northern counterscarp gallery and traverses on the covered way
Photo by John Weaver

At the gorge a sallyport gave access to the parade and was connected to the east counterscarp by a drawbridge. On each end of the gorge was a demibastion. Flank howitzers in the demibastions would provide flanking fire down the gorge. Loopholes in the parad- level wall of the gorge also provided protection.

Inside the main work on each end of the gorge are stairs that lead from the parade level down to the lower-tier rooms of the demibastions. Each demibastion contained a casemated gun room, quarters, and a magazine. Magazines of this period were lined with wood and had doors with brass hinges and narrow slits for ventilation. These

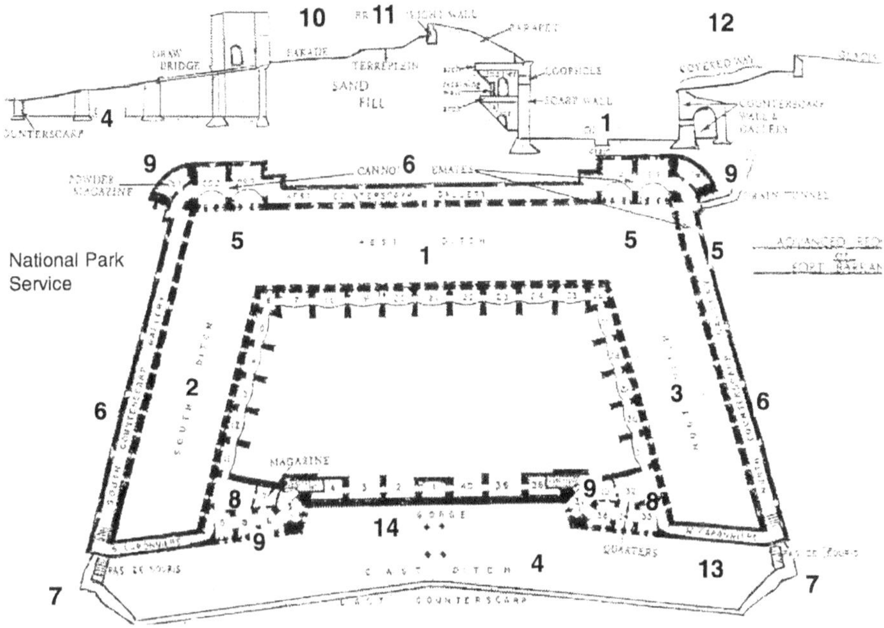

Plan legend: 1. West Ditch; 2. South Ditch; 3. North Ditch; 4. East Ditch; 5. Counterfire Rooms; 6. Counterscarp Galleries; 7. Pas De Souris; 8. Quarters; 9. Magazines; 10. Parad;e 11. Parapet; 12. Covered Way; 13. Tunnel 14. Gorge

Sallyport, drawbridge, and gorge demibastion Photo by John Weaver

measures were taken to keep the black powder dry and safe. A vaulted musket gallery runs around three sides of the main work on the lower-tier level. From the demibastions, subterranean tunnels ran to the counterscarp gallery on both sides.

Three sides of the counterscarp contained musket galleries. Each end of the west counterscarp contained two counterfire howitzers that could fire down the north and south ditches. Counterfire howitzers in the north gallery could sweep the west ditch. These counterfire rooms had ready powder magazines.

Arched recesses were located at regularly spaced intervals in the back wall of the counterscarp gallery. In the event of a siege, the brick in these recesses could be knocked down for the purpose of countermining. A mine could be dug under enemy positions and a powder charge set off or the mines could be used as a listening post to detect enemy mining activity.

Sallyport in the gorge wall with drawbridge that is raised Photo by John Weaver

The gorge area of the Advanced Redoubt Photo by John Weaver

During the Civil War, the Advanced Redoubt was occupied by Confederates. The Confederates used the Redoubt as an ordnance depot. After the Confederate evacuation of Pensacola on May 9th, 1862, the post was occupied by the Union Army for the duration of the war. The fort suffered damage from retreating Confederates. The only shots fired in anger from the redoubt occurred during a brief skirmish with Alabama calvary in 1863.

After the war, the Redoubt was abandoned for over sixty years. The WPA attempted to restore it during the 1930s. Most of the restoration work was done in the mid-1970s by the National Park Service. Today the Advanced Redoubt is administered by the Gulf Islands National Seashore.

a) Inside and b) outside the gorge demibastion Photo by John Weaver

Defenses of Tampa Bay Florida

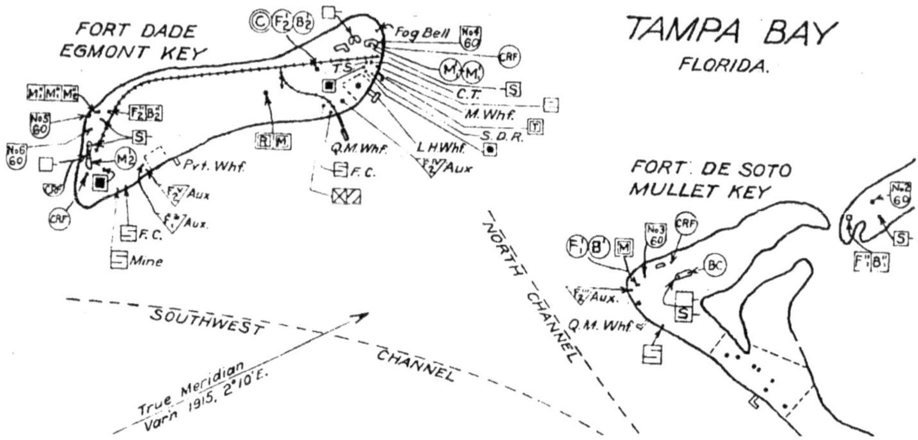

Fort De Soto, Mullet Key, Tampa Bay Begun 1898

In February, 1849, the schooner *Phoenix* laid anchor near the mouth of Tampa Bay. On board were four Army engineers that had come to survey Passage, Egmont, and Mullet Keys as possible sites for fortifications. One of the engineers was a young Brevet Colonel named Robert E. Lee. In March, 1849, they recommended Egmont and Mullet Keys for military use. The fortification sites they recommended would not be utilized until the Spanish-American War of 1898.

In 1898, Tampa became a major embarkation port for Brigadier General William Shafter's Fifth Corps. Twenty-five thousand men were encamped outside of Tampa and among them was the flamboyant leader of the Rough Riders, Lieutenant Colonel Theodore Roosevelt. Because of Tampa Bay's proximity to Cuba, it needed to be guarded.

Construction on Mullet Key began in November, 1898. A wharf, office, mess hall, workmen's quarters, and a stable were constructed during the first six months. A narrow-gauge railway was laid between the wharf and construction sites. In March, 1899, the foundation for the mortar battery was poured.

The post on Mullet Key was named for the Spanish explorer who had landed nearby in 1539, Hernando DeSoto. The Endicott fortification would contain a two-pit mortar battery and a two-gun 3-inch rapid-fire battery. At the time of construction, Fort DeSoto was a sub-post of Fort Dade on nearby Egmont Key.

The mortar battery was completed in May, 1900, containing eight 12-inch mortars. The steel rifled mortars were Model Ml890-Ml that had been manufactured by the Watervliet Arsenal in New York. The carriages, Model 1896-Ml, had been manufactured by the American Hoist and Derrick Company of St. Paul, Minnesota. Together, each mortar and carriage weighed 138,000 pounds.

The battery was named Battery Laidley in honor of Colonel Theodore T. S. Laidley, a veteran of the War with Mexico and the Civil War, who had died at Palatka, Florida. The battery contained powder and shell magazines, two observer stations, firing rooms, a relocating room, dynamo room, fuse storage rooms, and data booths. To-

One of the mortar pits in Battery Laidley with two 12-inch mortars
Mark Berhow Collection

Mortar data booth at Fort De Soto) Mark Berhow Collection

The ruins of the almost submerged Battery Bigelow Dale Manuel Collection

day, Battery Laidley still contains four of the only existing 12-inch mortars in North America.

In 1902, construction was completed on a battery with two 3-inch rapid-fire rifles. The battery was named Battery Bigelow in honor of First Lieutenant Aaron Bigelow, who was killed at the Battle of Lundy's Lane during the War of 1812. Battery Bigelow guarded the minefield in the north channel between Mullet and Egmont Keys. The 3-inch rifles were Driggs-Seabury Model 1898, on masking- parapet mounts. The projectiles weighed 15 pounds. A five-pound nitrocellulose powder charge propelled the projectile. At a maximum elevation of 12 degrees the range was 4.5 miles. Today, this battery in front of Battery Laidley is almost submerged in the surf.

Twenty-nine buildings were constructed on the post between 1900 and 1906. These buildings included a barracks, a hospital, a stable, a guardhouse, a blacksmith and carpenter shop, an administration office, a mess hall and kitchen, a bakery, and a storehouse. The buildings were wood frame with slate roofs.

In 1900, Fort De Soto was garrisoned by Company A, First Artillery. In April, 1907, the First Company was replaced by the 39th Company, Coast Artillery, which stayed until 1910. Florida State Troops often participated in training exercises at the post. Duty at the post was marred by horrendous swarms of bloodthirsty mosquitoes. In 1910, the Fort DeSoto garrison was transferred to Fort Morgan, Alabama, and the fort became inactive.

During WW I, four of the 12-inch mortars were shipped to Fort Rosecrans in San Diego, California. Through most of the war the detachment at Fort DeSoto had only two officers, two noncommissioned officers, and twenty-two privates.

In May, 1923, both Fort De Soto and Fort Dade were abandoned and the posts were put on a caretaker status. In 1926, both forts were declared surplus. In September, 1938, Pinellas County bought the 271 acre site.

In 1940, the Army used Mullet Key as a bombing range. The old post became a sub-post of MacDill Field in June, 1941. In August, 1949, Mullet Key was sold back to Pinellas County. On May 11, 1963, Fort DeSoto Park was officially dedicated.

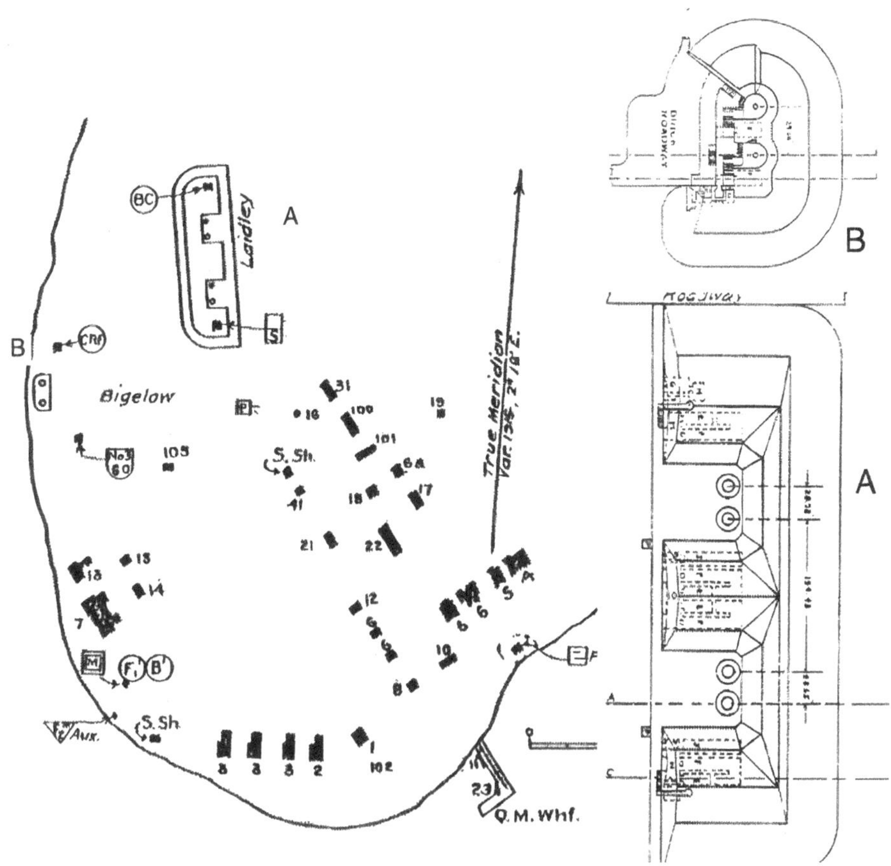

Plan of Fort De Soto: A. Battery Laidley; B. Battery Bigelow
Dale Manuel Collection

Fort Dade, Egmont Key, Tampa Bay Begun 1898

Fort Dade was established during the Spanish-American War of 1898. Egmont Key was selected because its position commanded both the southwest and north channels. Endicott-period batteries would be constructed on the north and south ends of the island to guard both channels. The fort was named in honor of Major Francis L. Dade, who, along with most of his command, was massacred by Seminoles in 1835.

Batteries Page and Burchsted were constructed on the south end of the island. Built in 1898, Battery Burchsted mounted two 6-inch Armstrong rapid-fire rifles on pedestal mounts and one 3-inch rapid-fire rifle. Battery John Page mounted two 3-inch rapid-fire rifles. Later, an electric power plant was located between the batteries to supply them both. The two 6-inch Armstrong rifles remained in Battery Burchsted until 1982 when they were removed and restored. The 6-inch rifles are now on display at Fort DeSoto Park.

Three concrete batteries were constructed on the north end to guard the north channel. These batteries were named McIntosh, Howard, and Mellon. Completed in April, 1900, Battery McIntosh mounted two 8-inch rifles on disappearing carriages.

Lighthouse (Fort Dade) Dale Manuel Collection

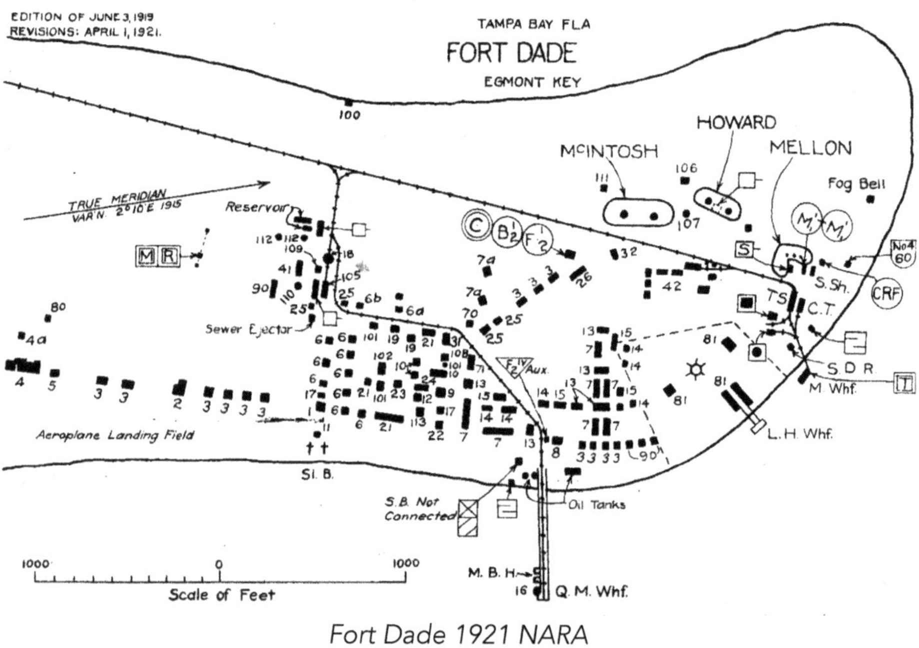

Fort Dade 1921 NARA

Battery McIntosh Mark Berhow Collection

Battery Mellon once held two 8-inch rifles and was rebuilt to hold three 3-inch rapid-fire rifles to guard the North Channel minefields Mark Berhow Collection

Battery Howard's damaged parapet and center structure are shown Mark Berhow Collection

The battery was named for Colonel James S. McIntosh who was killed at the Battle of Molino del Rey, Mexico. Slightly north of Battery McIntosh is Battery Howard. Battery Howard was constructed in 1906 and mounted two 6-inch rifles on disappearing carriages. The battery was named for Major Guy Howard, who was killed in the Philippines on October 22, 1899. Because of manpower shortages, there was no gunnery practice until August, 1916. During WW I, the guns were removed and shipped to France. Because of beach erosion, the concrete parapet of the battery was undermined and collapsed in a storm in 1995. The most northern battery is Battery Mellon, which was constructed in 1898 during the Spanish-American War and mounted two 8-inch rifles on modified Rodman carriages. In 1901, the battery was completely rebuilt to mount three 3-inch rapid-fire rifles for the protection of the north channel minefields. In 1920, the 3-inch rifles were removed. In 1943, the Navy built a wooden Harbor Command Post on top of the battery.

From 1898 to 1921, the Army planted two lines of mines in each shipping channel totaling 84 mines. Both ends of the island contained mine casemates and mine primary stations. The north end, near Battery Mellon, contained a mine storage building, cable tank building, loading room, dynamite storage building, and the mine wharf.

Mine Wharf Mark Berhow Collection

A 6-inch Armstrong rifle from Battery Buchsted Mark Berhow Collection

A narrow-gauge railroad ran from the quartermaster's wharf on the north channel to the batteries on the south and north ends. The railroad also ran to the mine center and engineer's wharf on the north channel.

Fort Dade became a city of approximately 300 people. The post had a hospital, post office, Y.M.C.A., power plant, telephone system, sewer system, bakery, fire department, tennis court, guard house, airplane runway, etc. Along with nearby Fort DeSoto, Fort Dade was put on caretaker status in 1923 and declared surplus in 1926.

In WW II, the post was used by the Coast Guard as a coast-watching station and training base. Part of the island is used as a pilot station by the Tampa Bay Pilot Association, part of the island is a Florida State Park, and part of the island is a wildlife sanctuary. In the century that the fort had been in existence, the island has been eroding at an alarming rate. The remains of the south batteries are now in the Gulf of Mexico, while the north batteries that were once 100 yards from the beach are now on or near the beach.

Defenses of the Dry Tortugas Florida

Fort Jefferson, Dry Tortugas Begun 1846

Fort Jefferson has been alternately called the Gibraltar of the Gulf or the Devil's Island of the Gulf; both comparisons are fitting. Located on Garden Key, which is 68 miles west of Key West in the Dry Tortugas, one can marvel at how so much material was imported to such an isolated location in the days before modern transportation. Ponce De Leon discovered and named these small sand-and-coral islands the Tortugas, or turtles, in 1513. The name Dry was added later to denote the islands had no fresh water.

Fort Jefferson is an enormous irregular hexagon that covers most of the 16-acre island. Built at a deepwater anchorage to guard the main sea lanes between New Orleans and the East Coast, it also kept the Dry Tortugas from being used as a base for invasion of the mainland.

Sketch of Fort Jefferson under construction

Fort Jefferson from the air. The fort is a gargantuan hexagon
Photo by John Weaver

This was one of the largest Third System forts ever constructed, with a half-mile perimeter. Fort Jefferson was built to fight an armada of warships. Construction began in 1846. In order to construct the foundations, Army Engineer Lieutenant Horatio Wright constructed cofferdams and drained them with windmills and steam-powered pumps. In the cofferdam, a foundation of concrete that measured 2 feet x 14 feet was poured. Work on the fort dragged on for about 30 years and it was never completed.

The fort had three tiers of guns behind its brick and concrete scarp that rises to fifty feet. The first two tiers had vaulted casemates. At each salient of the hexagon was a massive tower bastion. Each tower has 13 vaulted casemates. On each of the two flanks of the bastions there are six howitzer embrasures.

At each face of the bastions on the second tier is a single gun embrasure. Guns were mounted en barbette on the third tier. All the howitzers on the flanks of the six tower bastions were to provide flanking fire on the curtains of each front of the fort. In the north bastion a bakery contained two large ovens heated by a lower fire box.

On the parade side at each salient are spiral stair towers that climb to the top tiers. On the third tier terreplein are batteries for guns en barbette. The batteries are divided by sixteen traverse magazines. Iron catwalks on the inner side of the parade wall provided access to the magazines. The terreplein was completed in 1864.

The scarp walls taper from ten feet thick at the base to eight feet thick at the cordon. Just below the cordon is a decorative brick corbel series of arches that surrounds the entire work. At the embrasures, the thickness of the scarp is reduced to five feet.

National Park Service photo

There are 159 casemates in each of the first and second tiers. In addition to the gun positions in the casemates there are magazines and storerooms. Under the floors of 109 of the lower casemates are cisterns that hold about 11,000 gallons of water. Rainwater was collected on the terreplein and piped to the cisterns below.

The second-tier embrasures were never installed to limit the weight of the structure. In order to make these areas look complete, the open spaces were bricked in with one layer of brick. Most of this brickwork is now gone.

When construction started, it was believed the island was on a solid coral reef. It was learned later the fort's foundations were built on sand and coral boulders. The tremendous weight of the fort has caused sinking and resulting cracking of the scarp walls and cisterns.

On the parade was an assortment of buildings. A massive barrel-vaulted magazine located near the northeast bastion was begun in 1866 but never finished. A smaller magazine located near the southwest bastion had a double barrel vault. It was also begun in 1866 and never finished. The original Garden Key Lighthouse, built in 1825, was also located on the parade. In 1876, the present iron lighthouse on the terreplein replaced the older one. The top of this lighthouse stands 82 feet above the parade. Located on the parade's east side was the enlisted men's quarters, completed in 1875. Its dimensions were 35 feet x 338 feet and provided barracks for 1,000 men. There were verandas on each of the three stories and there was a detached kitchen for each of the ten companies the barracks would hold. This barracks burned in 1912.

Another large barracks was the Officers' Quarters located in the north portion of the parade. Completed in 1874, its dimensions were 44 feet x 286 feet. It had 63

View of terreplein with a Parrott rifle Photo by John Weaver

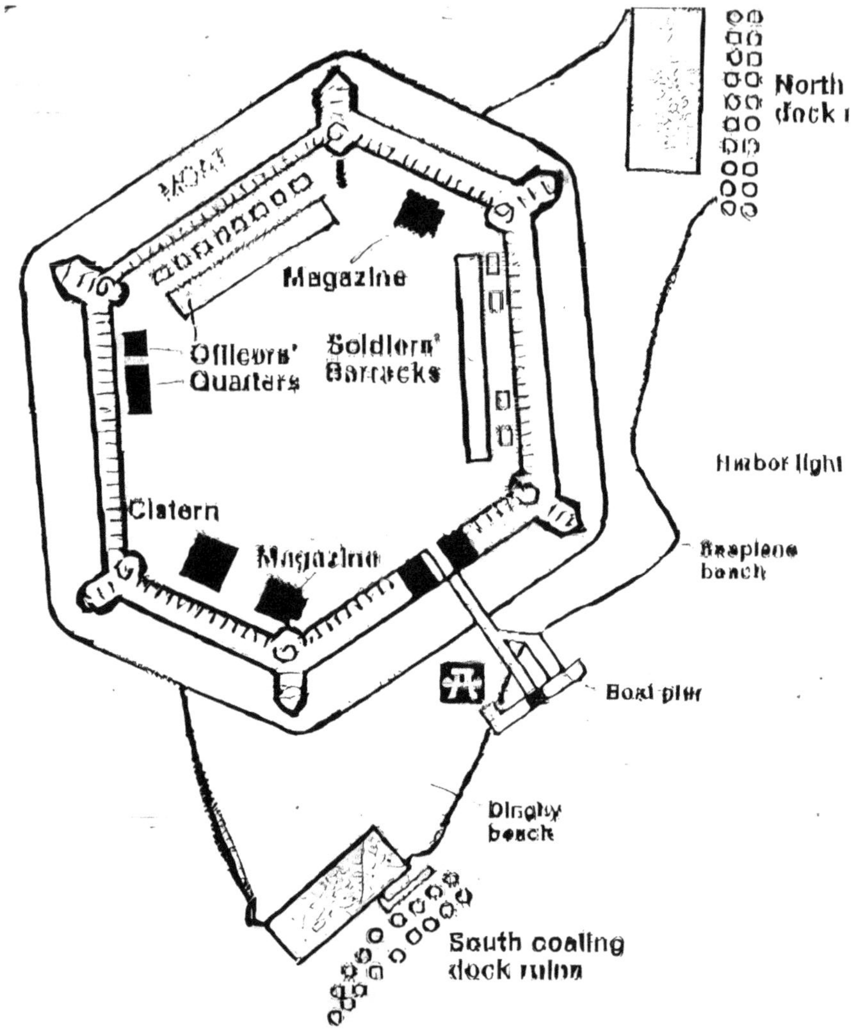

Fort Jefferson is a six-sided irregular polygon. National Park Service.

rooms and was three stories tall, but the detached kitchens in back were only two stories tall. In the northwestern corner of the parade are three two-story buildings, completed in 1866. These buildings served as the engineers', clerks', and doctors' quarters. An officers' quarters was to have been built and the three buildings were to be converted into kitchens, but the quarters were never constructed. Also on the parade is a massive hot-shot furnace and a 94,000 gallon cistern.

When the Civil War started, the fort was only half-completed and was unarmed. The fort was hurriedly occupied by Union troops on January 19, 1861. The wartime garrison present was about 500 men. In 1863, armament for the barbette tier consisted of six 10-inch columbiads. Ten years later these were replaced by fifteen-inch Rodman smoothbore guns and 8-inch Parrott rifles. In 1865, so-called "Lincoln Conspirators" Samuel Arnold, Michael O'Loughlin, Edward Spangler, and Doctor Samuel A. Mudd were incarcerated at Fort Jefferson.

The iron lighthouse that was constructed in 1876 to replace the original Garden Key lighthouse Photo by John Weaver

The sallyport and drawbridge are the entrance to the fort Photo by John Weaver

The massive tower bastions are the largest ever constructed in the U.S. Photo by John Weaver

A large hot-shot furnace is on one corner of the parade Photo by John Weaver

The curtain gallery of the fort shows its massive scale and length
Photo by John Weaver

Dr. Mudd and Fort Jefferson

Doctor Mudd had been sentenced by a military panel to life at hard labor for setting the broken leg of Lincoln assassin John Wilkes Booth. In 1867, a yellow fever epidemic caused 270 of 300 men to fall ill. One of the first of the 38 fatalities was the post doctor, Major Joseph Smith. Doctor Mudd worked day and night to fight the disease. Two years later the doctor was pardoned by President Johnson in the last days of his presidency. Dr. Mudd left Fort Jefferson on March 8, 1869.

The parade of Fort Jefferson Photo by John Weaver

After another outbreak of yellow fever and a disastrous hurricane, Fort Jefferson was abandoned by the Army in 1874 and construction stopped the following year. The Dry Tortugas served as a quarantine station between 1888 and 1900. In 1898, the fort was garrisoned again by artillery troops and the Navy used the fort as a coaling station. In January, 1898, the battleship Maine left the Dry Tortugas for the last time on her ill-fated voyage to Havana, Cuba. On February 15, 1898, the Maine was blown up in Havana Harbor, sparking the Spanish-American War.

One of the Navy's first wireless stations was set up at Fort Jefferson in 1904. The fort served as a seaplane base in World War I. In 1935, President Franklin D. Roosevelt proclaimed the Dry Tortugas a National Monument.

Defenses of Key West Florida

Fort Taylor, Key West Begun 1846

In 1846, construction began on Forts Taylor and Jefferson. Fort Taylor was named for Zachary Taylor, hero of the Mexican War and President of the United States.

The board of engineers selected the site for Fort Taylor during the coastal survey of 1822. A shallow shoal one thousand feet offshore from the western tip of Key West was chosen as the building site. The fort was built to protect the deep harbor of Key West with its navy squadron and strategic position in the Florida Strait. Commodore David Porter had used the island as a base in 1822 in his expulsion of pirates from Florida. The Fort was built on an artificial island of rock and sand fill.

The casemates were built on a grillage of timbers laid on the shoal. The foundation of the scarp was built of large granite blocks laid on a rock fill. On this granite foundation brick was laid. The fort's parade was just five and a half feet above high water level.

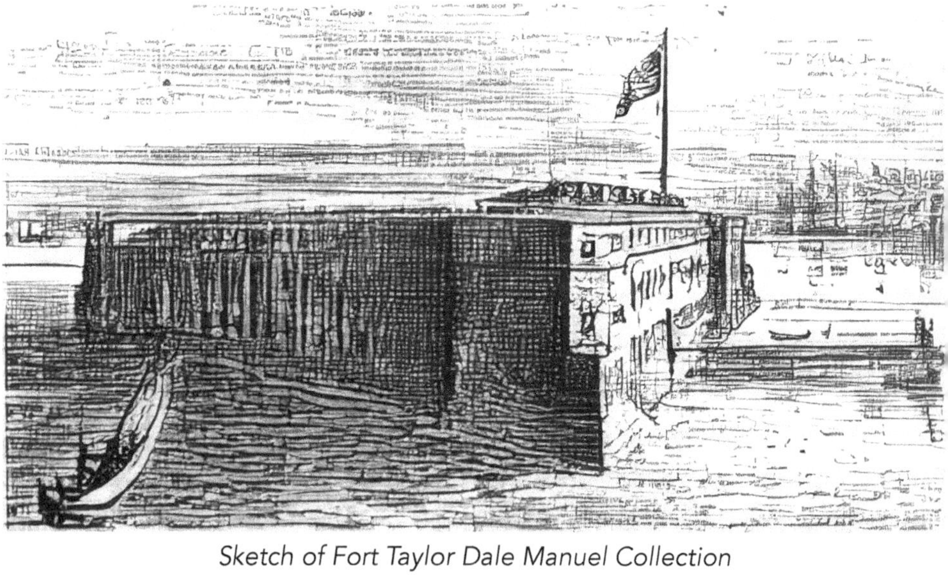

Sketch of Fort Taylor Dale Manuel Collection

The fort is a trapezoid in the shape of half a hexagon. The fort's long gorge measures 494 feet, each flank curtain 335 fee,t and its sea face 254 feet. At each of the four salients was a tower bastion. The original fort was three tiers high. On the gorge side was a three-story barracks. The barracks measured 56 feet high.

Fort Taylor as it now looks from the air. The fort was once 1000 feet offshore and three tiers high. Now the fort is landlocked with only one tier.
Dale Manuel Collection

A view from the parade of the north casemates with a stair tower at left
Photo by John Weaver.

Barracks at Fort Taylor were once three stories high and ran the length of the gorge side of the parade Photo by John Weaver

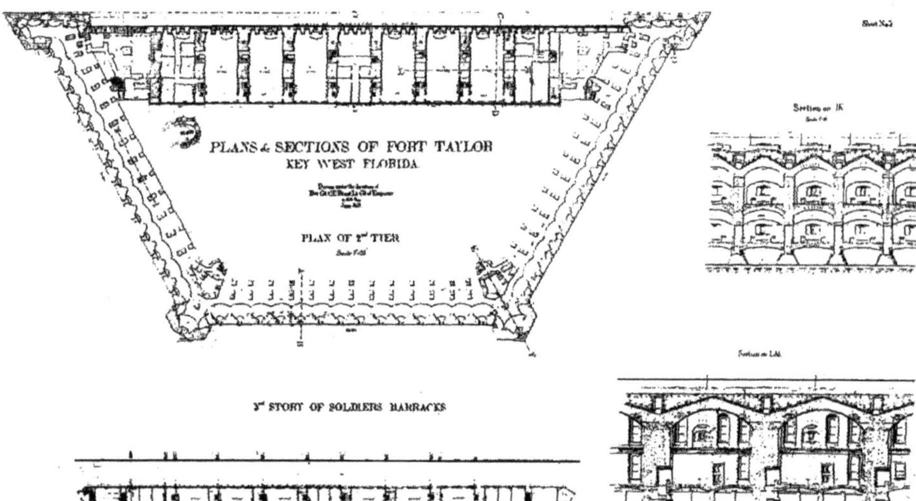

Plans of Fort Taylor. NARA.

Seth Eastman Painting of Fort Taylor. NARA.

The first and second tiers had vaulted casemates. Embrasures in the casemates still contain iron shutters invented by the fort's designer, Joseph Totten. The third tier's terreplein mounted guns en barbette. From the water at high tide to the parapet measured 45 feet.

One of the gun platforms of Battery Osceola Mark Berhow Collection

Fill with embedded guns on the south parapet Mark Berhow Collection

At each end of the gorge was a tower demibastion of three tiers while the other two salients sported a full tower bastion each. Both demibastions guarded the sally-port and gorge wall. The tower bastions could sweep the curtains of the primary and two secondary seacoast fronts that commanded the water approaches. The gorge faced the land side. The two full bastions contained a magazine and a spiral stair tower. Cupolas once covered each of the stair towers. Each end of the gorge also had stair towers that led up to the upper tiers.

Battery Adair. Both batteries were built inside the old fort. Mark Berhow Collection

The recovered gun carriages of Battery Adair Mark Berhow Collection

In the middle of the gorge scarp was the sallyport, the only entrance to the fort. It had a drawbridge with a mechanism to raise and lower it. The drawbridge connected to a causeway that ran one thousand feet to a coverface and on to the island.

Cisterns were built underneath the first-tier casemates and barracks. The largest cisterns were under the kitchens of the barracks. The system of collecting rainwater was augmented by the addition of a desalination plant in the early days of the fort. The three-story barracks contained quarters, offices, kitchens, and washrooms. Iron beams and vaulting were used between floors for support. All down the gorge were musket loopholes that also served as windows. The top-story doors of the barracks opened on the gorge terreplein that mounted guns en barbette. The fort's planned armament consisted of 106 guns and howitzers in the first- and second-tier casemates. Thirty-six more guns were to be mounted en barbette on the terreplein.

During the Civil War, Key West was headquarters for the Gulf Blockading Squadron. Key West was a secessionist city, controlled by the Union Army. In July, 1862, four hundred forty eight troops were garrisoned in the area. During the war the masts of hundreds of Confederate blockade runners could be seen for miles while detained under the guns of Fort Taylor.

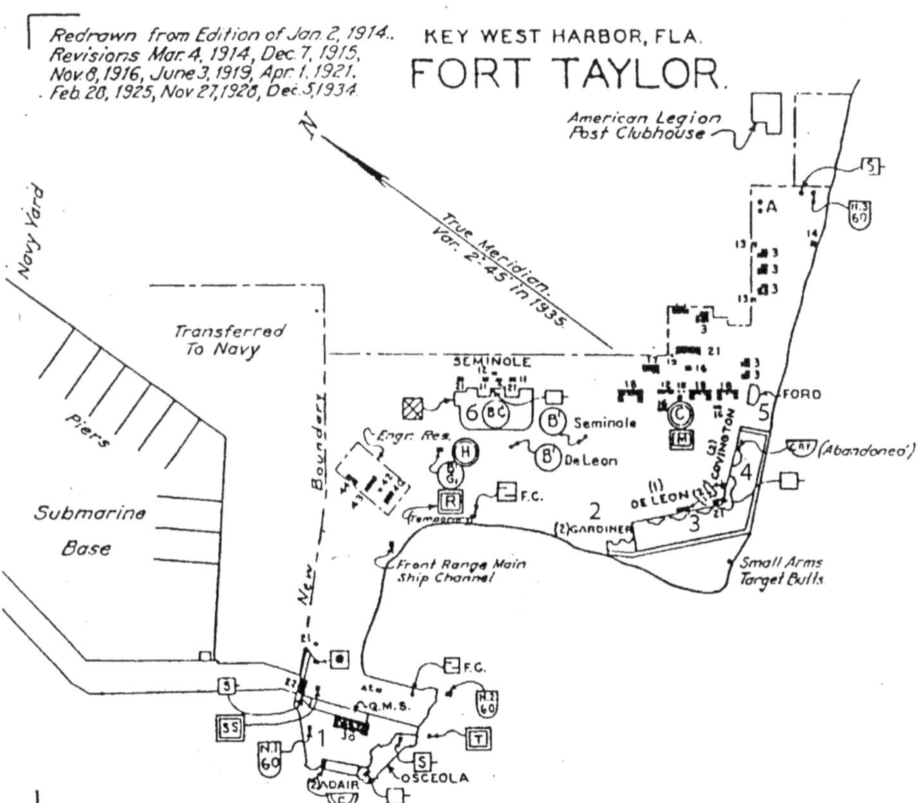

Plan of the Fort Taylor Reservation. 1. Old Fort; 2. Battery Gardiner; 3. Battery DeLeon; 4. Battery Covington; 5. Battery Ford; 6. Battery Seminole NARA

During the Spanish American War of 1898, the two top tiers of the casemates and barracks were torn down to make the fort less vulnerable. The first tier of casemates on the west and south curtains were filled in with sand and debris. Two large concrete batteries and magazines were built behind the south and west curtains. Battery Osceola contained two 12-inch rifles on barbette carriages. Built behind the west curtain, Battery Adair contained four 3-inch rapid-fire rifles. The 3-inch rifles were removed in 1920. The 12-inch rifles were removed in 1942 and shipped to England for its defense.

At the rear of the old fort was the harbor mine center with mine loading room, mine storage, cable tank, a service dynamite room, and a 1,223-foot mine wharf and tramway. The old mine casemate was built on top the old fort's coverface. Another mine casemate was built later next to Battery Seminole.

Other Endicott batteries were built on the nearby fort reservation between 1898 and 1904. Battery Seminole, a two-pit mortar battery, contained eight 12-inch mortars. Battery DeLeon mounted four 10-inch disappearing rifles. Battery Covington held two 8-inch disappearing rifles. Battery Gardiner mounted two 4.7-inch rapid-fire rifles. Battery Ford held two 3-inch rapid-fire rifles. Battery Dilworth mounted two 3-inch rapid-fire rifles. Battery De Kalb held two 6-inch rifles on barbette carriages. Most of these batteries have been demolished over the years. Only Batteries Adair, Osceola, Seminole and #231 have survived on the cramped Navy Base. Other island batteries that have survived are Batteries Inman inside the West Martello Tower and #232 at the Salt Ponds.

In World War II, a battery of four 90mm Anti-Motor-Torpedo-Boat dual-purpose guns was erected at the old fort. Two were fixed and two were mobile. One of the 90mm guns was mounted on top of the northwest corner of the fort. Searchlights were mounted on each side of the old fort. Near Battery Gardiner, construction began on a two 6-inch gun Battery Construction #231, but never armed or named. Construction began on a two 6-inch gun Battery Construction #232 located on Salt Pond Island between the two Martello Towers. Four 155mm guns on Panama mounts were constructed on top of Batteries DeLeon and Covington. Five 3-inch anti-aircraft guns were mounted near Battery Ford.

The fort once offshore is now landlocked. In the mid 1960s the surrounding area was dredged and the land was filled by the Navy. Fort Taylor became a Florida State Park after it was transferred to the state in 1976.

Battery Osceola Dale Manuel Collection

West Martello Tower, Key West Begun 1862

The construction of five towers around Key West was first suggested in the 1840s by the chief engineer constructing Fort Taylor, Captain George Dutton. The towers were to guard the beaches east of Fort Taylor from amphibious assault. Captain Dutton's recommendations were shelved because the Army favored more elaborate forts.

When the Civil War started in 1861, the Army found itself in possession of Fort Taylor in the secessionist town of Key West. Fort Taylor was vulnerable to assault from the island, since an attacking force could land on the beaches and march across the island to the fort. The urgencies and necessities of war caused the construction of two towers on Key West's south shore beaches, the most accessible landing locations. The twin towers were 1½ and 3½ miles from Fort Taylor. The tower closest to Fort Taylor was the West Martello Tower and the one farthest away was the East Martello Tower.

General Joseph G. Totten, one of the finest military engineers America has ever produced, designed the towers. Though called Martello towers, the central gun towers were actually designed after the tours-modeles favored by Napoleon I for coastal defense in France. Similar gun towers were built at Proctor's Landing, Louisiana, at Fort Winthrop, Massachusetts, and at Fort Porter, New York.

At the center of the fort was the central gun tower or citadel built of brick and trimmed with granite. The tower was 56 feet square and reached a height of 36 feet. The tower had three tiers with a square stairwell in the center containing an iron spiral staircase of 48 steps which led to the second and barbette tiers. The stairwell

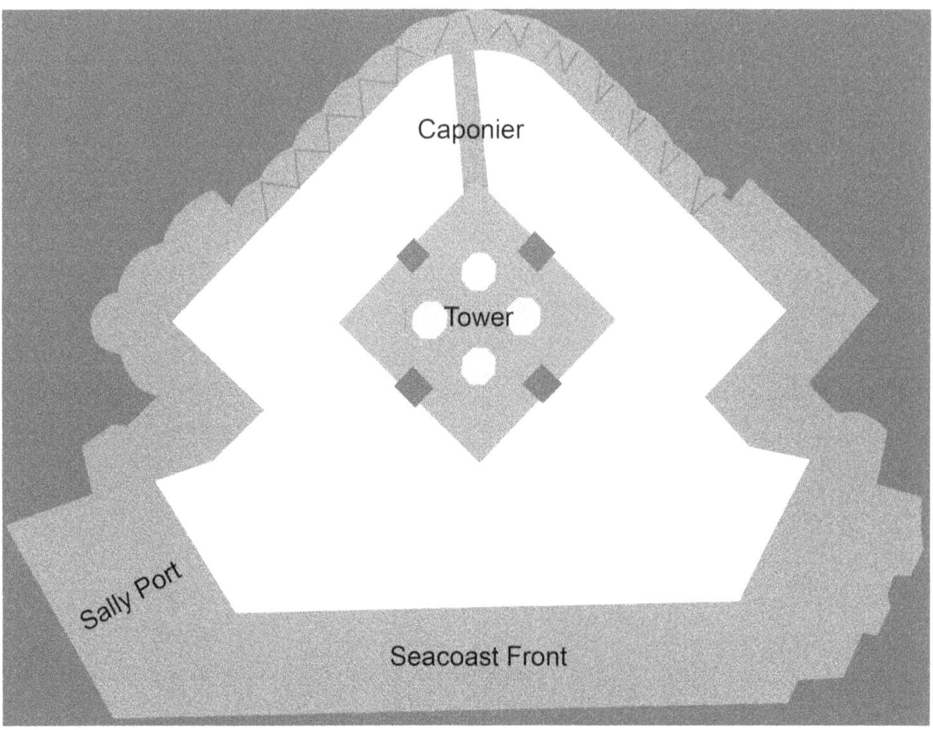

Plan of West Martello Tower John Weaver Collection

was covered on top by a wooden cupola. A granite machicoulis protruded out of the barbette tier in the center of each side of the tower. From this machicoulis, muskets could be fired down or hollow projectiles dropped on enemy infantry below. The first and second tiers had loopholes for muskets. These tiers contained barracks, a kitchen, a mess, officers' quarters, and magazines. The top tier had four circular emplacements for guns mounted en barbette on center-pintle carriages.

Ruins of the scarp of the West Martello Tower Photo by John Weaver

The sea battery scarp Photo by John Weaver

Remaining vaulted arch in the ruins of the old central tower
Photo by John Weaver

Inside a casemate of the northeast corner of the counterscarp gallery. In the center is one of the counterfire embrasures. Mark Berhow Collection

The tower was encircled by a dry ditch with a masonry counterscarp. To flank the tower's sides on the land side, a counterscarp gallery with musket loopholes for counterfire was built. The original fort had a 12-foot-high earth-covered glacis on the seaward side.

As the design of the fort evolved, casemated batteries of two tiers were to have been added to the seaward side. Only one tier was completed. This design alteration was to provide the firepower of 28 guns against any naval attack. A wet ditch 8 feet wide by 6 feet deep was to front the battery and was to have wrapped around to the counterscarp. This ditch was not constructed.

Work on the fort was often interrupted by lack of material, labor shortages, and yellow fever epidemics. Although the fort was mostly complete, the second-tier casemates of the seaward battery were never finished. Work stopped in 1866.

As the years passed after the Civil War, the West Tower began to deteriorate. Gunners from Fort Taylor used the structure for target practice. Further demolition and modifications occurred over the years to support other coastal defenses. The fort served as a lookout post during the Spanish American War. In 1904, an Endicott-period battery was built behind the seaward side casemates and was named Battery Inman. The battery held two 3-inch rapid-fire rifles. A steel rangefinder station was built on top of the remains of the old central brick tower. The battery was named in honor of

A view of the counterscarp gallery Photo by John Weaver

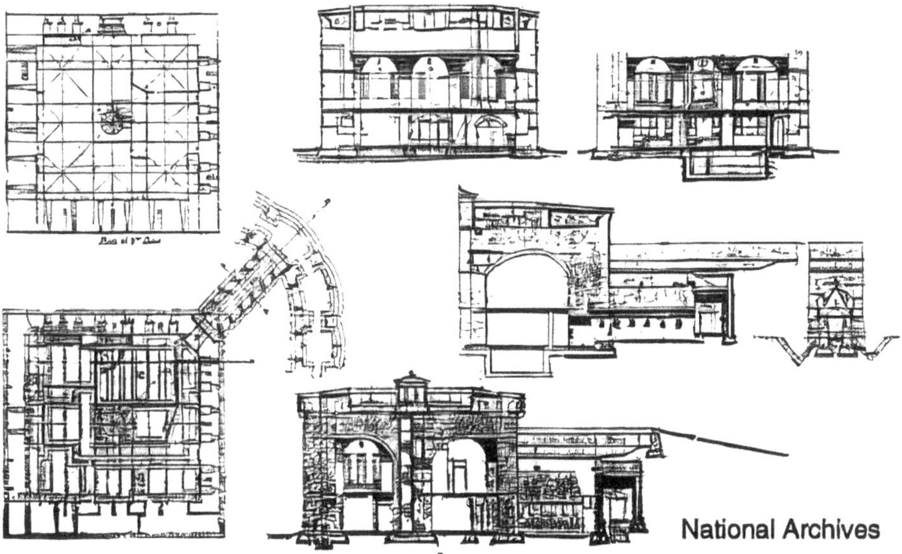

Plan of tower and original fort configuration without the seaside battery NARA

Captain Shadrach Inman who was killed in action at Musgrove's Mill, South Carolina, on August 18, 1780.

During World War II, four 90mm Anti-Motor-Torpedo-Boat dual-purpose gun batteries were located near the fort, two fixed and two mobile. Nearby temporary barracks housed the troops. Sixty-inch searchlight stations were built on each side of the fort. After the war the property was sold as surplus.

One of two 3-inch gun emplacements for Battery Inman Mark Berhow Collection

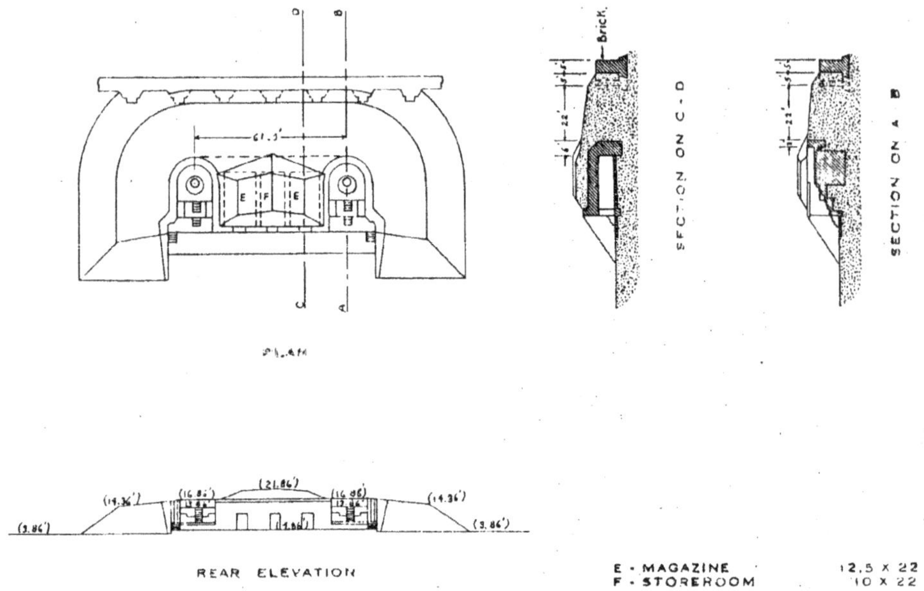

Plan and section of Battery Inman NARA

The fort is open to the public and operated by the Key West Garden Club. Most of the fort is in ruins. The northeast corner of the counterscarp galleries is the only portion of the fort that is intact. This, incidentally, is the side farthest away from the incoming shells that were fired from Fort Taylor. Only a small portion of the central tower's first-tier walls and central stairwell survive. The western side that faced Fort Taylor is open between the casemate battery wing and the service sallyport of the counterscarp. The seaward-side casemated battery scarp has been battered down to the level of the bottom of the embrasure sills.

East Martello Tower, Key West Begun 1862

The twin of the West Martello Tower is the East Martello Tower, which is considerably more intact than its sister fort. Built around its central three-tiered gun tower, it has a counterscarp gallery that has five casemated gun rooms on each end. The tower was to protect the eastern extremity of the island from amphibious assault. A narrow-gauge railroad ran from a wharf near Fort Taylor to both towers to bring in supplies and large quantities of brick, granite, and other materials.

The central gun tower, or citadel, like the similar tours-modeles tower at Proctor's Landing, Louisiana, was one of the first to use iron girders along with barrel-and-groin vaulted arches. All these structures were necessary to support the tremendous weight of four guns on the barbette tier. Though never installed, the tower was to have supported four 100-pounder Parrott rifles. The weight of the tube of each of these guns was 9,700 lbs. The weight of the guns, wrought iron carriages, and force exerted while firing would have to be structurally supported.

East Martello Tower exterior Photo by John Weaver

East Martello Tower as seen from the air. Up front is the casemated battery and in the center is the tower. Whitt Collection

An embrasure of a sea front casemate Photo by John Weaver

Inside the gallery of the seaside water battery Photo by John Weaver

Interior of the second tier of the central gun tower Photo by John Weaver

Exterior of the central gun tower. The white object on the top is a machicoulis Photo by John Weaver.

The counterscarp galleries were lined with musket loopholes. The sallyport in the counterscarp was used to bring in material during construction. It was to have been bricked up and covered with earth when the fort was finished. The horizontal arches on the outside of the counterscarp were to buttress it against the weight of the earthen glacis.

The only entrance to the fort from the outside was to have been from a drawbridge from the top of the glacis to a second-tier sallyport in the tower. The counterscarp gallery was only accessible from inside the base of the tower through a covered passageway. This covered passageway also had musket loopholes on both sides that served as a caponier for firing down the dry ditch.

In the event of infantry assault from the land side, the tower's second-tier sallyport could be closed by raising a drawbridge. As infantry progressed toward the tower, guns on the 36-foot-high tower would fire shot, then grapeshot, and finally canister. As infantry came nearer, riflemen from the second-tier musket loopholes of the tower would fire. Any infantry that dropped into the dry ditch for an attempt to scale the tower could be fired on with canister from counterfire rooms in the counterscarp gallery. Twenty-four-pounder flank howitzers would have the same effect as oversized shotguns. These casemated flank guns could sweep all four faces of the tower. Infantry in the ditch could be fired on by riflemen through loopholes in the first tier of the tower, in the counterscarp, and the caponier.

The fort's seaward-side casemated gun battery, like the West Tower, was added later to provide the overwhelming firepower of 28 guns against a naval attack. It was to have been two tiers high. The second tier was never finished.

The East Martello Tower is in excellent shape and is well maintained. It is operated as a museum and art gallery by the Key West Art and Historical Society.

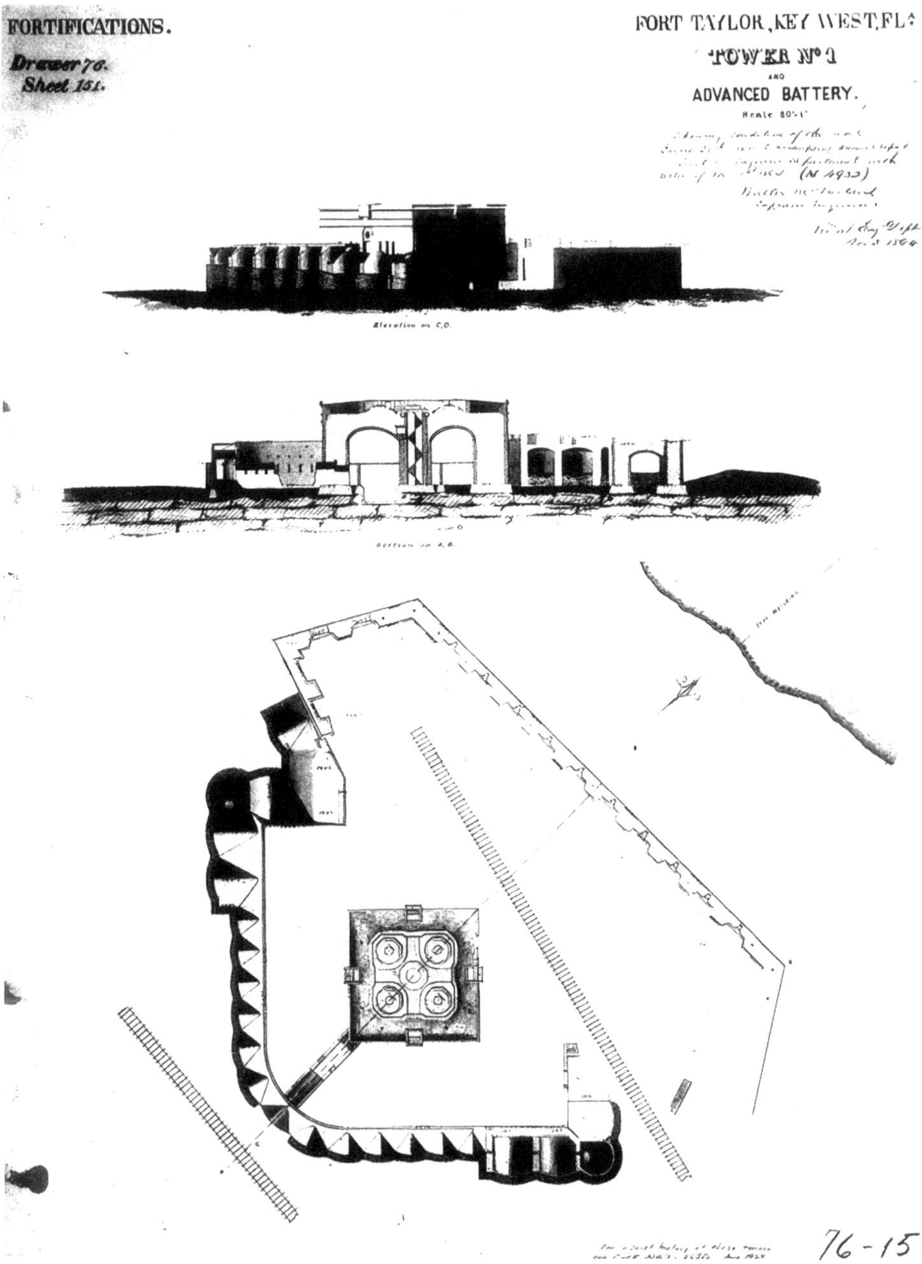
FORTIFICATIONS.
Drawer 76.
Sheet 151.
FORT TAYLOR, KEY WEST, FLA
TOWER No 1
AND
ADVANCED BATTERY.
(N 4932)
Elevation on C.D.
76-15

Parade of the East Key West tower. Central tower is to the right and the rear of the seacoast casemates is just left of the tower Photo by John Weaver

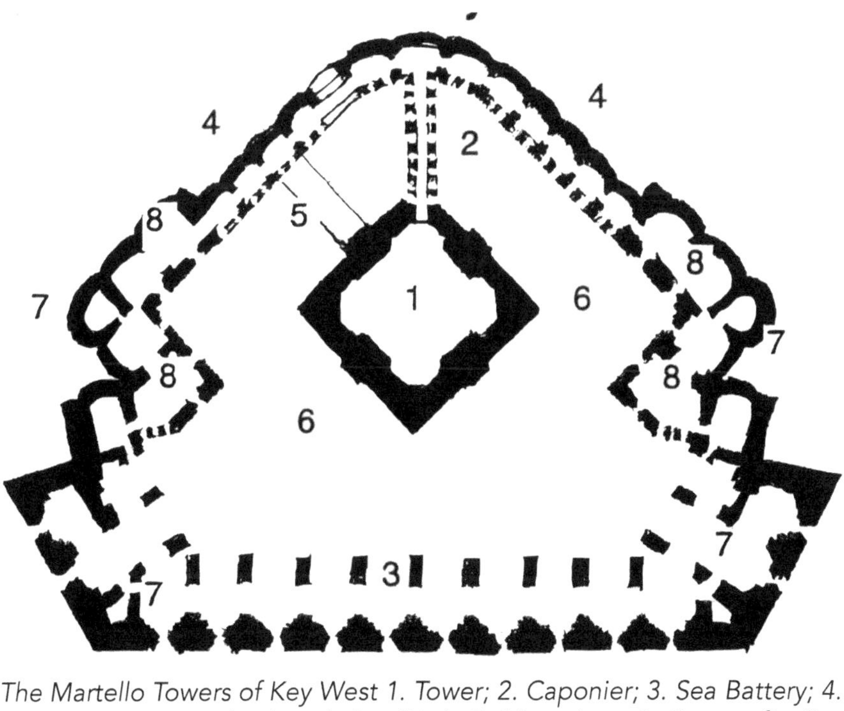

The Martello Towers of Key West 1. Tower; 2. Caponier; 3. Sea Battery; 4. Counterscarp; 5. Drawbridge; 6. Dry Ditch; 7. Magazines; 8. Counterfire Room Dale Manuel Collection

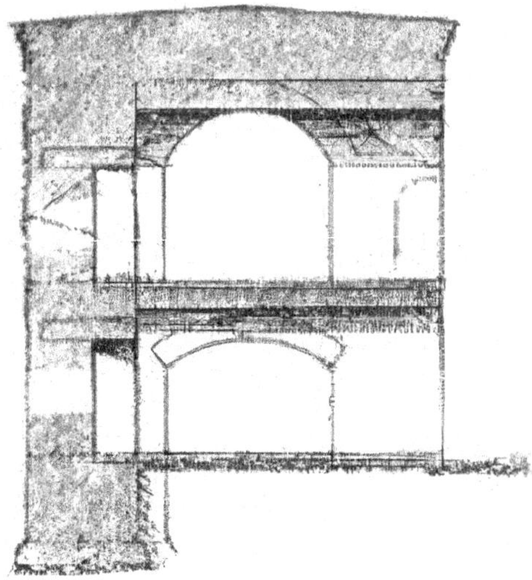

a-d Plans of the two-tiered sea front casemates. Only one tier was ever completed. Dale Manuel Collection

Corner of the counterscarp gallery that shows counterfire embrasures
Photo by John Weaver

Defenses of St. Augustine Florida

Castillo de San Marcos (Fort Marion) St. Augustine Begun 1672

The Castillo de San Marcos was begun by the Spanish in 1672 to replace a series of nine wooden forts that had been built at St. Augustine over the period of a century. St. Augustine had been established in 1565 and is the oldest European settlement in the United States. The Castillo was built not only to guard St. Augustine but to reinforce Spain's claim on Florida and protect Spanish commerce. St. Augustine occupied a strategic position near the Gulf Stream, used by the treasure fleets to cross the Atlantic to Spain.

The Castillo was built with a hard-shell rock called coquina, quarried on nearby Anastasia Island. The fort is a square with four bastions. Each side measures 320 feet, and the walls are 36 feet high. The fort was designed in the old-world style of fortifications by Spanish engineer Ignacio Daza. The terreplein is supported by barrel-vaulted bombproofs. Crenels in the parapets provided openings for cannon fire. The vaulted bombproofs were built between 1738-1742 and 1751-1756. A wet ditch, 42 feet wide, once surrounded the entire work.

The fort's only entrance to its square parade is a sallyport on the south side. The sallyport is connected to a ravelin by a drawbridge. This ravelin was built in 1762 and is a triangular defensive structure. Though never completed, the ravelin would have held cannon on its earth terreplein. Since the ravelin was built in the wet ditch, another drawbridge was used to connect it to the covered way. The covered way was built

Drawing of Fort Marion Castillo De San Marcos Dale Manuel Collection

between 1672 and 1762. A five-foot-high revetment of coquina separated the covered way from the earthen glacis. The glacis was constructed between 1672 and 1758.

In 1763, Florida was ceded to the British in exchange for Cuba, which the British had captured. Under the British the Castillo was renamed Fort St. Mark. In 1783, Florida was retroceded to Spain. In 1808, the Spanish built the Cuba line of defense as the northern city wall. The line was built of palm logs with a coquina city gate. The Castillo anchored the eastern end of this line.

Seafront battery with hot-shot furnace. In the background is a bastion with watch tower or bartizan. Photo by John Weaver

The parade of the square-shaped Castillo de San Marcos shows vaulted quarters and the door at right center is the post chapel. Photo by John Weaver

The old city gate was in the Cubo wall. The Castillo was the eastern terminus of this city defense. Dale Manuel Collection

The ravelin that connects two drawbridges Photo by John Weaver

In 1821, Spain ceded Florida to the United States. In 1825, the Army renamed the Castillo Fort Marion, in honor of Revolutionary War leader Francis Marion. In the years between 1821 and the Second Seminole War, 1833-1842, the Army had little interest in the fort because St. Augustine was not believed to be in a strategic position. During the war, the fort served as the base of operations. In 1837, Chiefs Osceola and Coacoochie of the Seminole nation were captured along with almost a hundred Indians, under a flag of truce. These Seminoles were held prisoners in Fort Marion. On the night of November 29, twenty of the Indians, including Coacoochie, escaped. Afterward, Osceola was moved to Fort Moultrie in Charleston Harbor.

Between 1833 and 1842, the present seawall was built on top of the remains of the old Spanish seawall. The eastern part of the wet ditch was filled in between 1842 and 1844 to build the Third System water battery. The battery is 41 feet wide by 248 feet long. Sixteen gun emplacements were built in this open battery. A hot-shot furnace was built in the center of the battery.

After the Second Seminole War, Fort Marion was gradually reduced caretaker status. On January 7, 1861, State of Florida troops seized the fort. Confederate occupation of Fort Marion was brief. When Union gunboats appeared off St. Augustine in March, 1862, Confederates evacuated the city. The Federal troops occupied the fort for the duration of the war.

In the mid-1870s, the fort was used as a prison for the detention of members of the western tribes. Lieutenant Richard H. Pratt attempted to assimilate and educate the Indians during their three-year stay at the fort. In 1880, Pratt established the Carlisle Indian Training School in Pennsylvania. In April, 1886, 77 Apaches were impris-

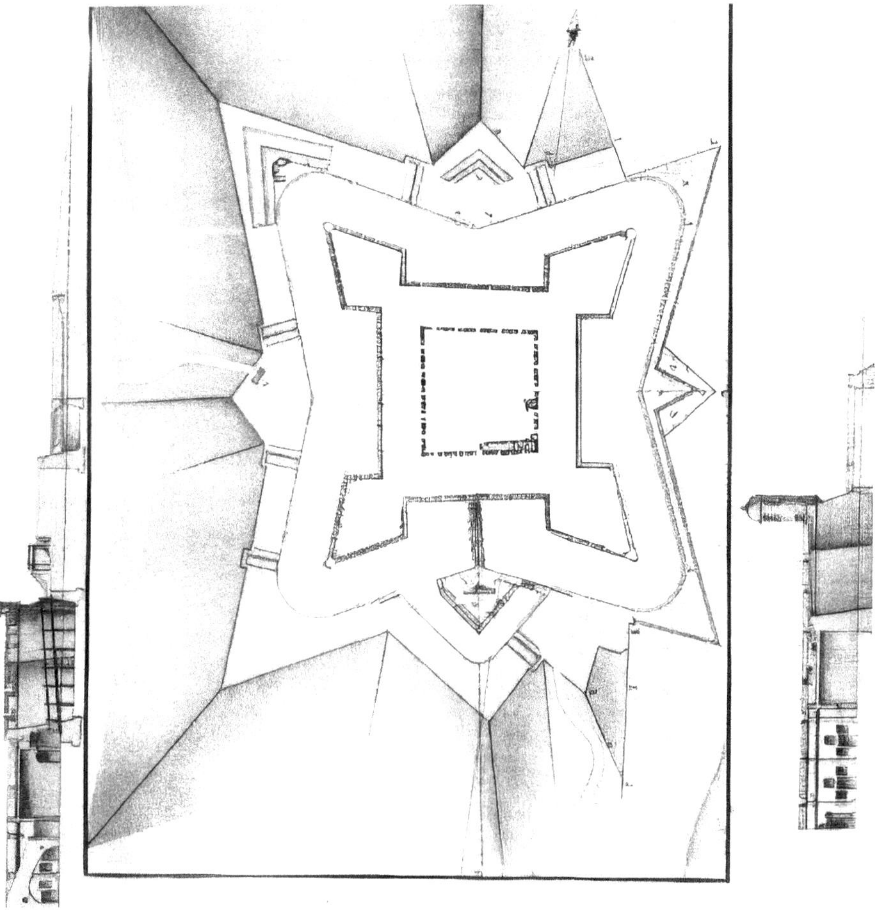

Plan of Castillo de San Marcos. NARA

oned at Fort Marion. By January, 1887, the total number of captives at the fort was 447. A special school was set up to educate this group as well. In 1887, the prisoners were taken to reservations in Oklahoma.

Efforts at reconstruction and repair occurred in 1884, when President Chester A. Arthur appropriated $5,000. This was a precedent for the expenditure of federal funds for the preservation of historical structures. After the Spanish-American War, in 1900 the fort was decommissioned. In 1924, President Calvin Coolidge declared Fort Marion and nearby Fort Matanzas National Monuments. In 1942, Congress restored the original name of Castillo de San Marcos to the fort. The fort has been fully restored and is operated by the National Park Service.

Fort Matanzas, Matanzas Inlet, St. Augustine Begun 1740

Though never used by the United States for coastal defense, Fort Matanzas, a small coquina tower, was a key to the defense of Spanish St. Augustine. Located at Matanzas Inlet, 14 miles south of St. Augustine, this fort guarded the back door to the city. The name Matanzas means "slaughters". The area is the site of the murder of hundreds of Frenchmen by the Spanish.

The French shipwrecked here during an attempt to capture St. Augustine in 1565. The fort was begun in 1740 and finished in 1742. The tower is three stories tall with

Sketch of Fort Matanzas Dale Manuel Collection

The small Fort Matanzas was built of coquina by the Spanish in 1740 Photo by John Weaver

an open battery platform and barracks on the first story. In the second story are barracks, storage rooms, and a magazine. The top of the tower has a musket platform. A cistern was in the base of the tower.

When the British gained control of Florida in 1763, they too garrisoned the tower with seven soldiers and two cannon. The tower was under British control until Florida was retroceded to Spain in 1783.

The tower deteriorated over the years. The fort was designated a National Monument in 1924. and was fully restored and operated by the National Park Service.

Defenses of St. Mary's River, Florida

Fort Clinch, Amelia Island, St. Mary's River Begun 1847

Fort Clinch was built on the northern tip of Amelia Island to protect the entrance to Cumberland Sound and the confluence of the Amelia and the St. Mary's Rivers. The fort guarded the strategic town of Fernandina, Florida. Fernandina was the terminus of a railroad that ran across Florida to Cedar Key on the Gulf of Mexico.

The fort is a pentagon with five tower bastions. The design is very similar to that of Fort Gaines, Alabama. Designed by General Joseph Totten, the design makes use of two technological advances. One is the tower bastion. The other development, originated by Napoleon's Minister of War, Lazare Carnot, was a detached scarp with musket loopholes. Behind the scarp was a narrow passage called a chemin de ronde, where riflemen could fire through the loopholes. Starting at this narrow passage the exterior slope of the earthen rampart rose to the superior slope and finally to the parapet that had a masonry revetment. Behind the parapet were guns mounted en barbette. The idea was that if the curtain, or scarp, of the Carnot wall was battered down or breached, the earthen rampart would remain intact.

Gun line at Fort Clinch Photo by John Weaver

The scarp of each tower bastion is higher than the scarp of the curtain. Each tower bastion had two tiers. The first tier's barrel-and-groin vaulted rooms contained four flank howitzer embrasures and four musket loopholes. Each tower contained a magazine. To collect rainwater, cisterns were below the floor of each bastion. A spiral stairway led up the tower to a gun mounted on the barbette tier. Each tower could sweep the faces and flanks on both sides.

On the fort's parade were barracks, a bakery, a store house, lumber sheds, a blacksmith's shop, officers' kitchens, and laundries. Enlisted men's barracks along the pa-

Enlisted Men's Barracks Photo by John Weaver

Guardhouse & Prison Photo by John Weaver

rade were never finished. An officers' quarters on the northeast side of the parade was also never completed. In the middle of each front was a gallery that led from the parade to the chemin de ronde. A gallery connected a latrine on the southeast front that was connected to the chemin de ronde by passages. Even the walls of the latrine had musket loopholes to fire on the slopes. A gallery ran from the parade leading to each bastion.

A dry ditch surrounded the work. A sallyport in the gorge wall was connected to a curved passageway through the glacis by a drawbridge. A brick sea wall along the north shore gave some protection from the waves. Jetties and a breakwater were later added. Fort Clinch was begun in 1847 and was never completed. The fort was named for General Duncan Lamont Clinch, who was a veteran of the War of 1812 and Florida's Seminole Wars.

In the back of a tower bastion Photo by John Weaver

The sallyport and gorge wall Photo by John Weaver

At the start of the Civil War, not one of the fort's designed 70 guns was mounted. Only the two bastions that faced the sound were complete. Confederate forces seized Fort Clinch in 1861.

With the Federal capture of many of the coastal islands along the Atlantic and the Gulf, it was realized that the South could not hold them without a strong navy. In early 1862, General Robert E. Lee ordered the withdrawal of forces on Amelia and Cumberland Islands. On February 28, 1862, a large Union flotilla sailed from newly captured Port Royal Sound, South Carolina. The vessels were mainly gunboats. The Confederates, upon receiving word of the approach of a superior force, began evacuating the fort and the town of Fernandina. On March 3, the evacuation was completed just in time, for Federal gunboats had just arrived. Fort Clinch was the first of the United States forts to be recaptured after being seized by the Confederates.

Carnot detached scarp, as seen from the ramparts Photo by John Weaver

Drawbridge, to the left of the picture, connects coverface with the sallyport. In the background is one of the gorge tower bastions. Photo by John Weaver

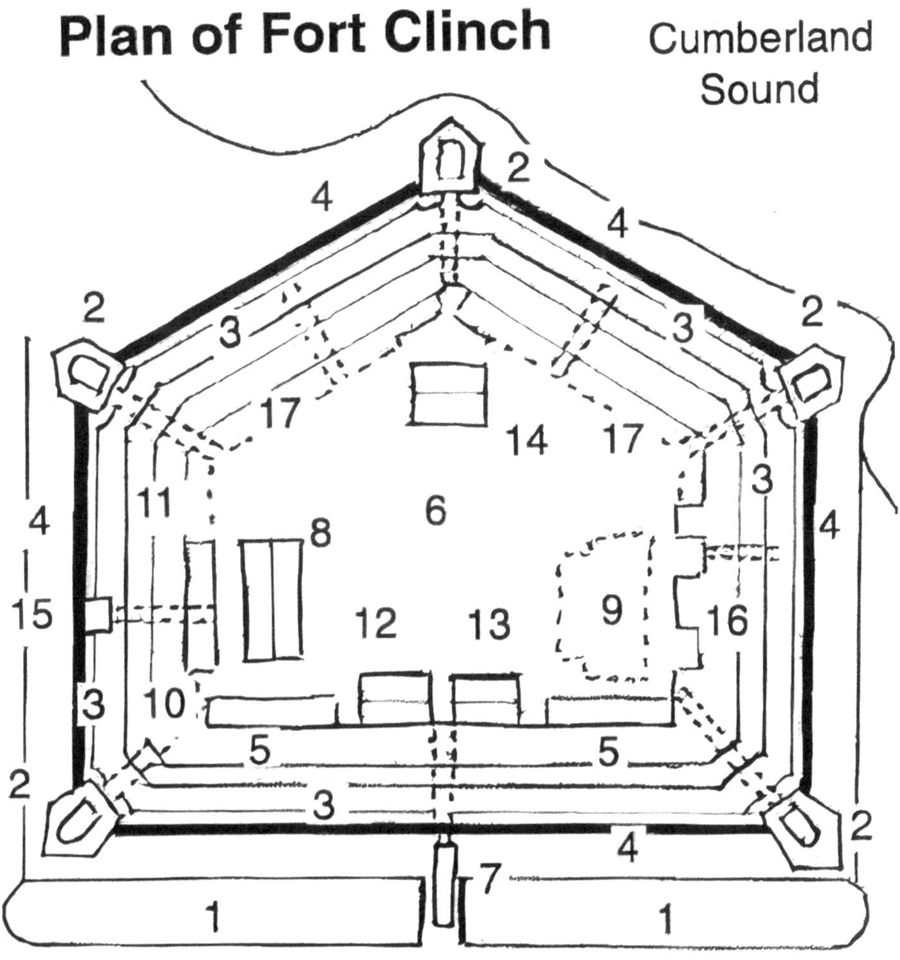

Plan of Fort Clinch. 1. Glacis; 2. Tower Bastions; 3. Chemin de Ronde; 4. Carnot Wall; 5. Ramps; 6. Parade; 7. Drawbridge; 8. Enlisted Barracks; 9. Unfinished Officers' Quarters; 10. Bakery; 11. Blacksmith; 12. Prison; 13. Guardhouse; 14. Q.M. Storeroom; 15. Latrine; 16. Kitchens and Laundries; 17. Parados (Unfinished) Dale Manuel Collection

About half of the construction of the fort occurred during the Civil War by Union troops. If you look at the fort's scarp, you can see a clear line of demarcation between a brownish-colored brick on the bottom and a cherry-red brick on the top. This line marks the progress of construction before the Civil War. After the war began, the local brownish brick was not available to Federal troops, so the red brick had to be obtained from Northern brickyards. In 1867, work on the fort was stopped.

In the years after the war, the fort was only maintained on a caretaker status until 1884. The fort was briefly activated in 1898 during the Spanish-American War. A small battery was erected on the rampart near the north tower. An 8-inch breech-load-

The parade, looking toward the north tower bastion and sound
Photo by John Weaver

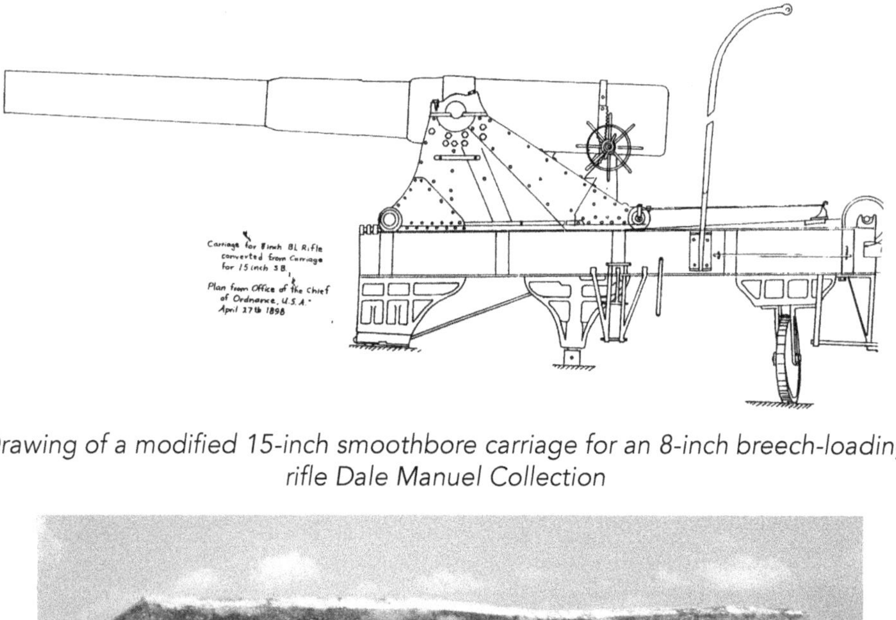

Drawing of a modified 15-inch smoothbore carriage for an 8-inch breech-loading rifle Dale Manuel Collection

This Endicott concrete shield and emplacement once held an 8-inch breech-loading rifle. Mark Berhow Collection

ing rifle was mounted on a modified 15-inch smoothbore carriage. The gun was removed in 1900. During WW II, the fort was used solely as a communications and lookout post.

In 1935, the State of Florida made the first purchase of the land for a state park that included the fort and would eventually cover most of the northern end of Amelia Island. During the 1930s, the Civilian Conservation Corps helped develop the park. The fort and park are well maintained and interpreted by Florida State Parks.

Defenses of Lower Mobile Bay

No matter what nation's flag has flown at Mobile, Alabama, whether French, English, Spanish, Confederate, or United States, the city's defense and that of its bay has always been considered of paramount concern. Mobile was not only a seaport, but a river port as well, with trade routes deep into the interior of the state. The navigable Alabama and Tombigbee Rivers converge above the city to form the Mobile River. In the 1850s the city became a railroad terminus.

Sketch of Mobile skyline Dale Manuel Collection

This prosperous city boasted a U.S. arsenal, shipyards, ship stores, shops, and other industries. Its main export was, of course, "King Cotton". Perhaps more importantly, the reason the United States was so concerned about the defense of the city and bay of the same name was because of its position between New Orleans and Pensacola. Here was the largest and finest harbor on the Gulf Coast between the South's largest and most strategic city at that time, New Orleans, and the large Navy Yard at Pensacola, with a fine harbor as well. The city and bay had to be guarded with the best defensive system available.

Fort Bowyer, Mobile Point Begun 1813

In 1813, Colonel John Bowyer led troops to establish a small sand-and-log fort on Mobile Point that would later bear the Colonel's name. During the War of 1812, General Andrew Jackson took great pains to strengthen Fort Bowyer. He was aware of its great importance, for this fort commanded the entrance to Mobile Bay. Major William Lawrence was sent to garrison the fort with 161 men.

On September 15, 1814, four British vessels of war, the Hermes, Caron, Sophia, and Anaconda sailed into the mouth of the bay and opened a bombardment on the fort. The squadron was led by Captain Henry Percy. Simultaneously, a landing party of 600 Indians and 130 Royal Marines, led by Colonel Edward Nicolls, attacked the fort from the land side. The garrison of the fort held firm. After three hours the Hermes was disabled by cannon fire from the fort and was blown up by her crew after drifting out of control. The British withdrew the next day.

After the British were foiled at the Battle of New Orleans, they knew that in order to conquer the Gulf Coast they must take the City of Mobile. They also knew that without taking Fort Bowyer they could not take Mobile. They planned a campaign against the fort by using Dauphin Island as a staging area. Thirty-eight Royal Navy

Drawing of casemate interior Dale Manuel Collection

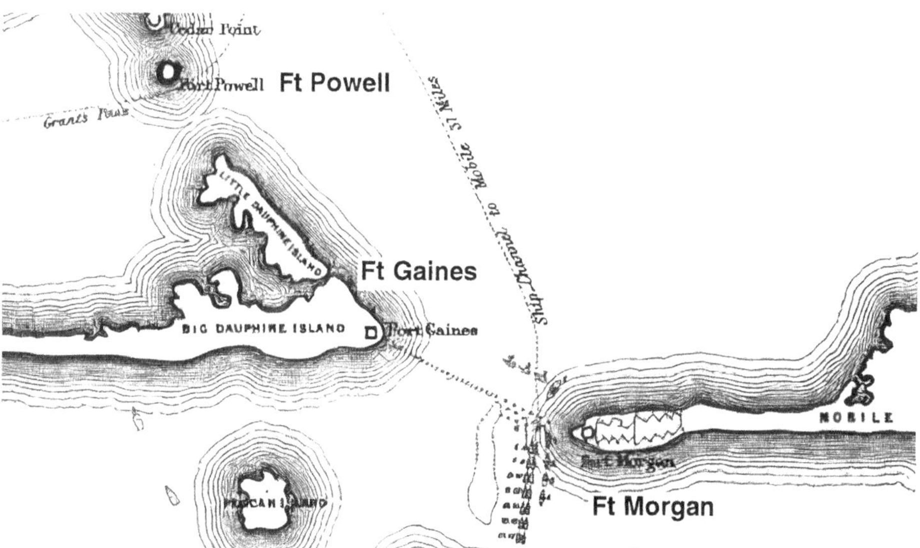

Map of the Confederate defenses of the Lower Mobile Bay
Dale Manuel Collection

vessels attacked the small fort from the north, west, and south. On February 8, 1815, the British landed 5,000 troops east of the fort on the bay side. A strong trench was dug across the peninsula and guns were emplaced. The beleaguered garrison of Americans were surrounded. Colonel Lawrence was in command of only 320 men and did what he could to defend his position. Also within the walls of the fort were 20 women and 16 children. On February 9, the besiegers opened approach trenches and parallels toward the fort. By the eleventh day they had dug to within forty yards of the fort.

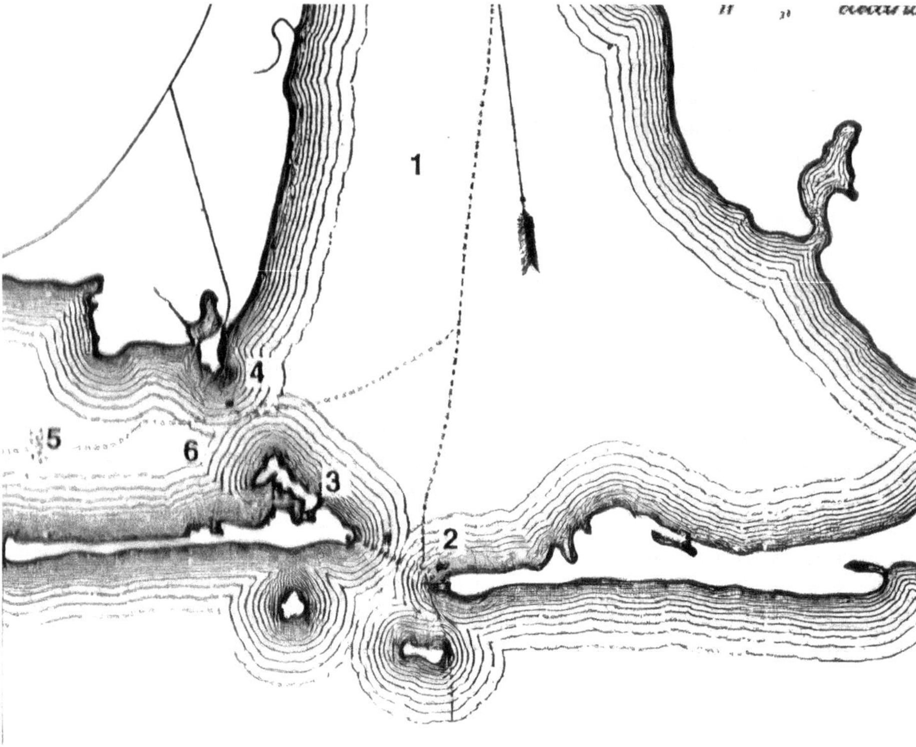

Map of Mobile Bay. Legend: 1. Mobile Bay; 2. Mobile Point; 3. Dauphin Island; 4. Cedar Point; 5. Mississippi Sound; 6. Grant's Pass Dale Manuel Collection

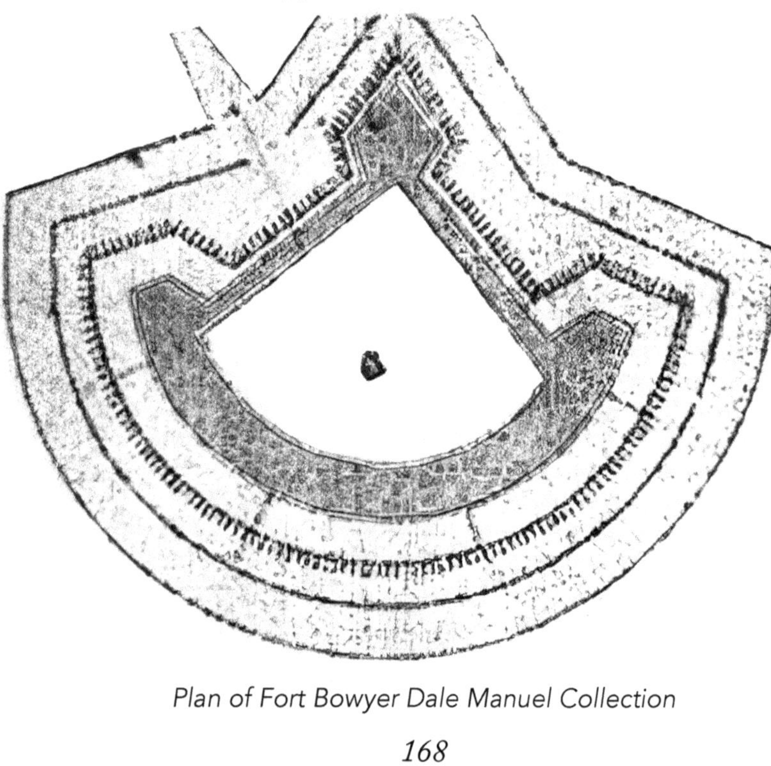

Plan of Fort Bowyer Dale Manuel Collection

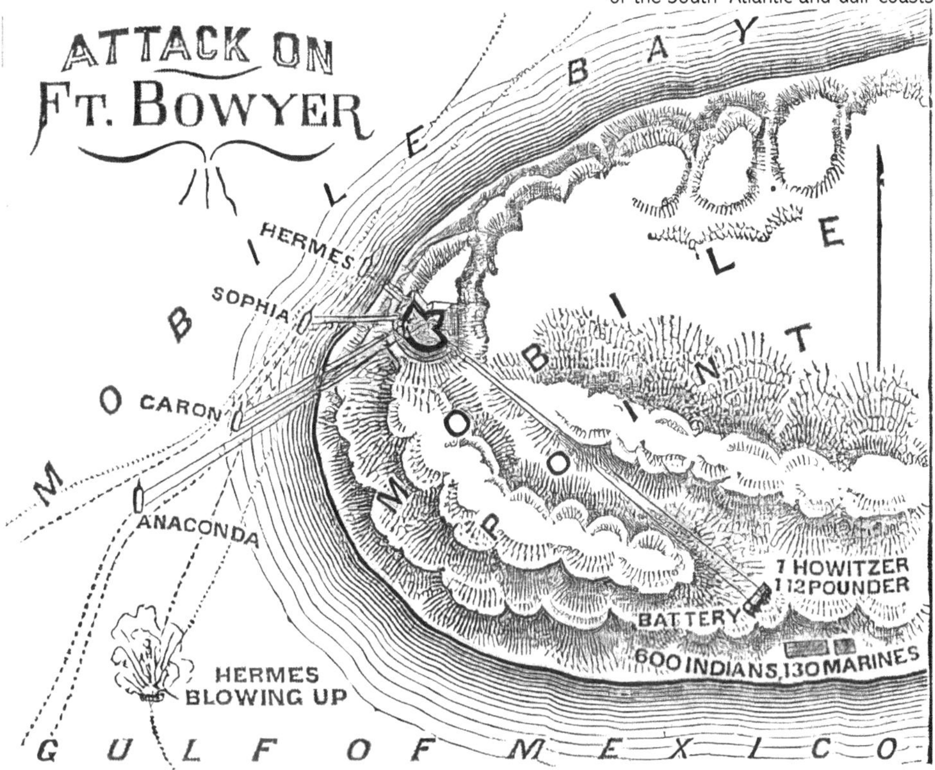

Map of the first battle of Fort Bowyer on September 15, 1814.
Dale Manuel Collection

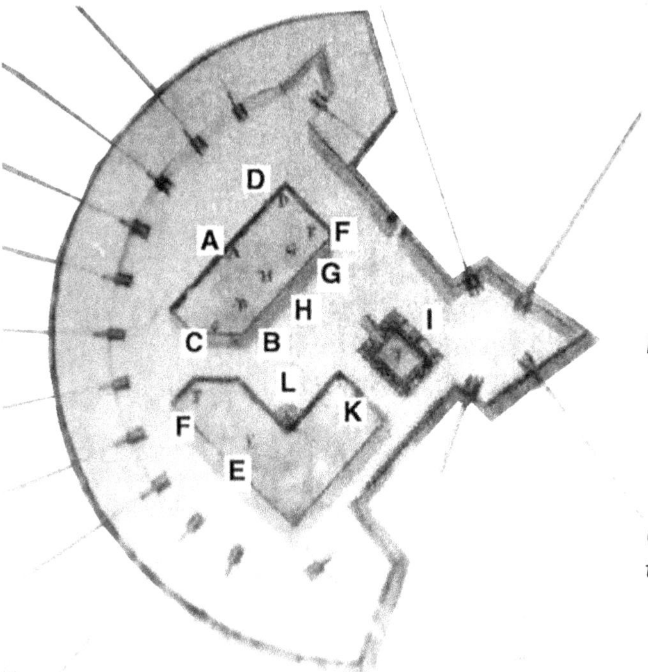

This plan of Fort Bowyer shows a semi-circular channel-side. Two demibastions and a full bastion gave flanking fire for land-side defense. A. Officers' Quarters; B. Surgeon's Quarters; 3. Pharmacy; D. Officers' Kitchens; E. Barracks; F. Guardhouse; G. Contractor's Store; H. Blacksmith; I. Powder Magazine; K. Privy; L. Cistern
Dale Manuel Collection

Lawrence surrendered to British General John Lambert at noon on February 11, 1815. Fort Bowyer had changed hands for only two days when word of the peace treaty reached the British. The British did not approach Mobile, for the battle for Fort Bowyer had been the last battle of the War of 1812. Fort Bowyer was destroyed by a storm in the 1820s and its site now lies in the present-day channel.

Tower at Pass Aux Herons Never Constructed

Once there was a mere speck of land midway between Cedar Point and Little Dauphin Island. This small oyster bank was not only mid-channel but was located between Pass aux Herons and the newer Grant's Pass. On this tiny island, the U.S. Army Corps of Engineers planned to construct a gun tower. The engineers named the island Tower Island. The tower was to guard the back door to Mobile Bay through Mississippi Sound.

For years the Corps considered a fortification for the old Pass aux Herons, but it was so shallow that only the lightest vessels of war could pass. Because of the shallow-

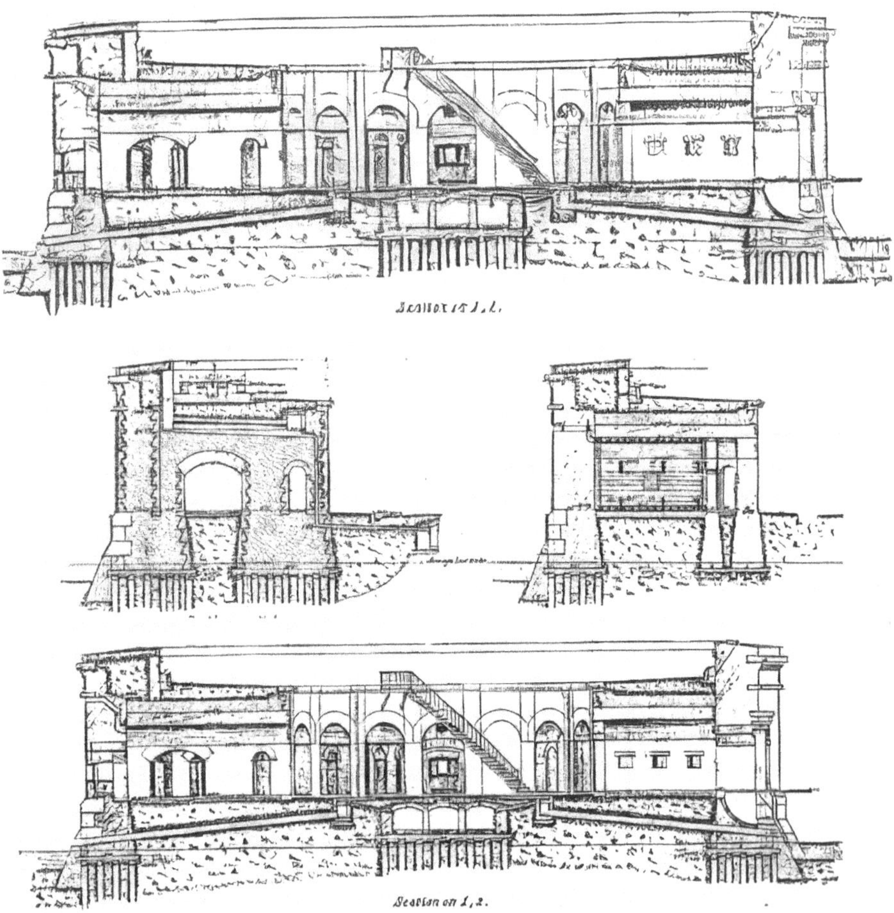

Section drawing of the Tower at Pass Aux Heron Dale Manuel Collection

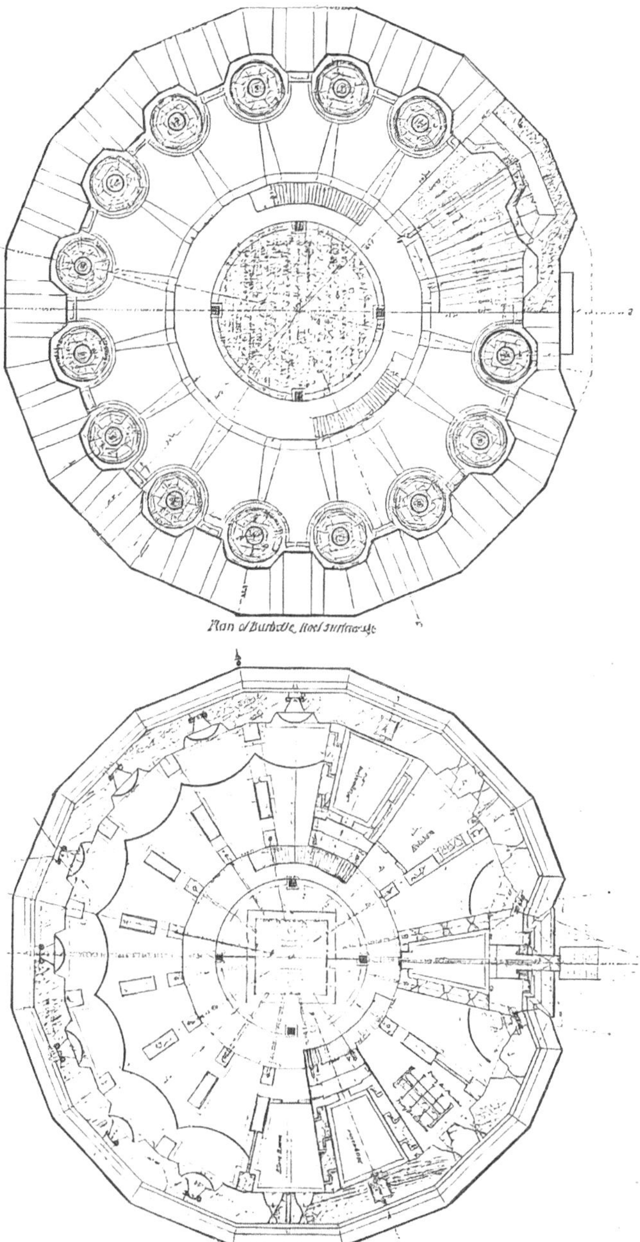

Sections of the gun tower Dale Manuel Collection

ness of the pass there was not that great of a concern, so the project was placed on a 5th class of construction priority. Two developments made the need for a fortification more urgent: the advent of steamships with shallow-draft keels and the dredging of Grant's Pass in 1839. Grant's Pass would become the main route from Mobile to New Orleans for passenger steamers and mail packets plying the Mississippi Sound. First

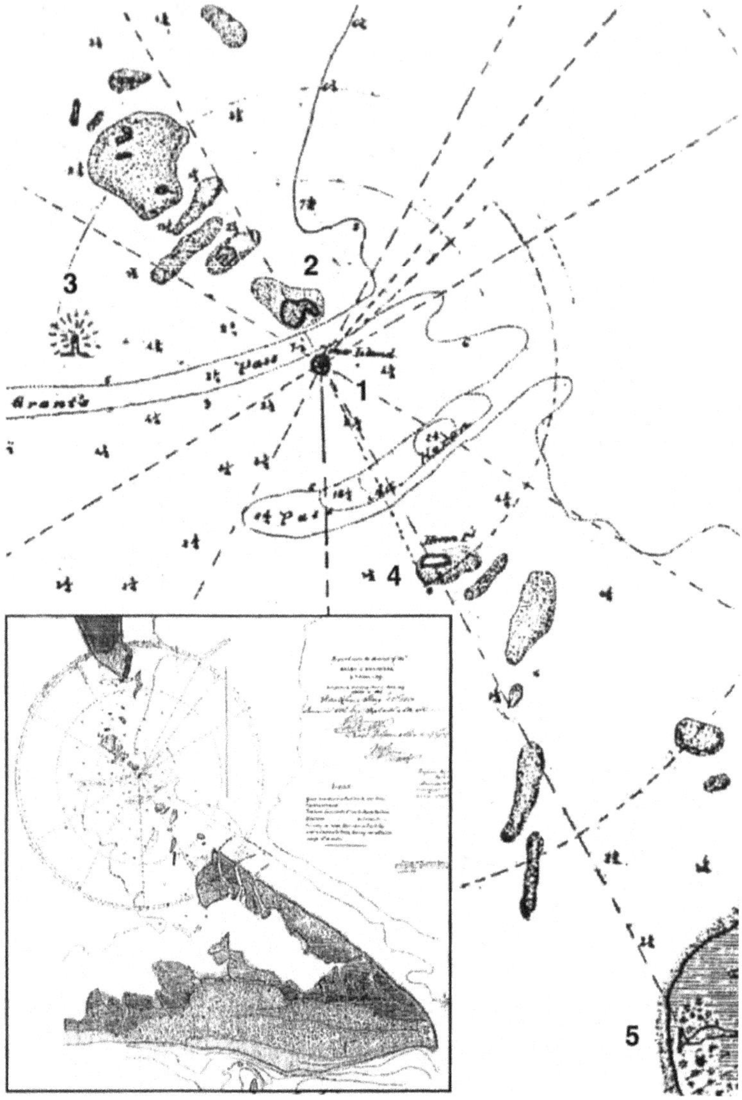

This 1851 chart of Grant's Pass and Pass Aux Herons shows the location of Tower Island and the toll station at Grant's Pass. The circular pattern shows the planned range of cannon fire from Tower Island. Legend: 1. Tower Island; 2. Toll station; 3. Lighthouse; 4. Heron Island; 5. Little Dauphin Island. Dale Manuel Collection

dredged by Captain John Grant with a dredge of his own invention, the pass had a tollhouse and private lighthouse on the north side of the channel.

By 1860, plans for the tower were nearing finalization, but the upcoming Civil War intervened, and construction never began. Had the tower been built, plans show that it would have been two-tiered with casemates and a barbette level. The foundation was to have been laid on 1,002 piles and a timber grillage.

With 16 faces the tower had the appearance of a circular form to give 360 degrees of fire. Fort Massachusetts on nearby Ship Island, Mississippi, has a similar design. As planned, the armament called for 16 barbette guns, eight casemate guns and two flank guns to guard the sallyport. The sallyport would have been entered from a drawbridge that connected a wharf. Loopholes would have lined the sallyport on both sides. Toward the gorge there were to be eight musket loopholes in the scarp. Vaulted casemates would have surrounded a small circular parade. Plans show cisterns under the parade floor and two iron stairs to climb to the barbette tier.

Casemates would have also contained two wood-lined magazines, a kitchen, a storeroom, and a divided privy. The perimeter of the work would have measured 400 feet. The estimated cost of the work was $122,285.20.

Although the United States would never construct the tower, the Confederate States would build a fort to guard the passes, called Fort Powell. Even though the location of Fort Powell has been reported as Tower Island, the actual site was probably the Grant's Pass toll and lighthouse keeper's station location on the north side of the channel. The private lighthouse was in advance of the station on the Mississippi Sound entrance to the pass. Most Civil War maps show the location of the fort on the north side of the channel. As can be seen from pre-Civil War charts, Tower Island is clearly on the south side of the channel.

The Sketchbook of Captain William S. Trask

Union officer Captain William S. Trask served with the Michigan 6th Heavy Artillery, Company H during the Civil War. He sketched scenes of his regiment's campaigns on the Mississippi River, in Louisiana, and of the Mobile Campaign. His company served on Dauphin Island during action against Fort Gaines. He drew scenes of Fort Powell, Fort Gaines, and Fort Morgan after the battle.

A drawing of the Grant's Pass lighthouse and toll station. Piles mark the pass. Dale Manuel Collection

Fort Morgan, Mobile Point Begun 1819

Fort Morgan was one of the early works of the Third, or Permanent, System to be started. Beginning in 1819, it was constructed near the site of the War of 1812 Fort Bowyer. Its location on Mobile Point is on the western end of a long, narrow peninsula that forms the eastern side of the mouth of Mobile Bay. Three miles across the mouth of the bay lies Dauphin Island, defining the western side of the mouth. Original plans called for twin forts to be constructed on these two points. This was to create a cross-

Photograph of the main sallyport of Fort Morgan as seen from inside the passage through the outworks. Beyond the main sallyport of the fort is a sallyport into the citadel. The inscription above the sallyport reads: Fort Morgan 1833. That is the year the fort was named in honor of General Daniel Morgan. John Weaver Collection

Photograph showing the damage inside one of the flank bastion casemates of Fort Morgan. Notice the holes in the casemate scarp, the damaged embrasure shutter, and the flank howitzers. Dale Manuel Collection

Fort Morgan's dry ditch and scarp of the main enceinte Photo by John Weaver

fire and adequately cover the wide entrance. Designed by General Simon Bernard in the finest traditions of the great French engineer Sabastien Vauban (1633-1707), the fort was a large bastioned pentagon. Each of the five bastions had two casemates in each flank to sweep the fort's curtains. Fort Morgan is one of the largest Gulf coast forts that was constructed. It shares a similar plan to Fort Jackson, Louisiana, but is larger and taller than that fort.

The fort was named in honor of Revolutionary War General Daniel Morgan. Morgan served in both the successful battle of Saratoga and the disastrous Quebec Campaign. Morgan also led a small force of militia and Continentals at the pivotal battle of the Cowpens on January 17, 1781. On the eve of the battle, Morgan assured his men that the "Old Wagoner" would crack his whip over "Benny." The next day the British regulars under Colonel Banastre Tarleton were routed by the Patriots.

Construction began in 1819 by private contractor. When two of the contractors died of yellow fever, Captain Rene De Russey of the Army Corps of Engineers recommended that construction be taken over by the Army Engineers in 1821. Brick was

The vaulted casemate galleries and stairs in one corner of the parade. Also visible is the outline of the citadel's inner wall and two cisterns. Photo by John Weaver

One of four large sallyport doors
Photo by John Weaver

A massive vaulted casemate gallery
Photo by John Weaver

made on the banks of the bay at the East Fish River, Red Bluffs, and on the Dog River. Most of the lime came from ancient Indian shell middens and lime kilns on Dauphin Island. The labor of about 3,000 slaves was used to construct the fort. The fort was built from the inside out. The citadel was constructed first, then the enceinte, and then outworks. The citadel was completed in 1825. The fort was considered finished by 1834 at a total cost of $1,026,777.41. After December, 1843, the fort was put on a caretaker status and was commanded by an ordnance sergeant.

The fort had a decagonal citadel, or defensive barracks, to serve as quarters and a place of last stand should the main enceinte be carried by an enemy. Fort Morgan's citadel was the largest constructed in the United States. The outside of the citadel had three tiers of musket loopholes, with the upper tier of loopholes commanding the barbette tier of the main work. The citadel's inner courtyard contained underground cisterns. The citadel's parade walls had doors that led to the barracks, sallyports that led to the outside, and musket loopholes that allowed fire on the courtyard. The citadel's inner wall also had ten sides. Each side contained either a door in the center or an arched gallery. Four large windows were also in each side, as well as two tiers of musket loopholes. There were as few as four loopholes or as many as twelve in a side. A veranda ran around the inner wall. The citadel had two stories of quarters and also contained baths and washrooms, kitchens, a mess hall, a hospital ward, and a dispensary.

The main work was two tiers high with casemates in the lower tier and a barbette level. Each casemate contained one gun embrasure and two musket loopholes to fire on the dry ditch. Each curtain contained seven casemates. The casemate's vaulting opened out to the parade. The casemates were connected by two parallel vaulted cur-

Loopholes that fired on the sallyport foyer area Photo by John Weaver

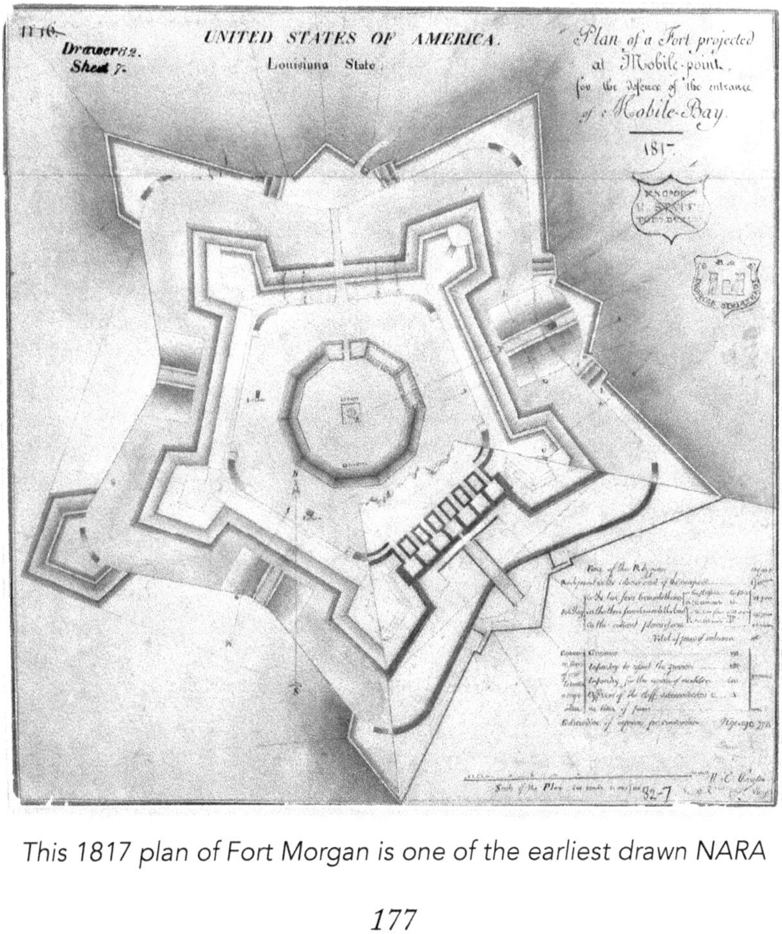

This 1817 plan of Fort Morgan is one of the earliest drawn NARA

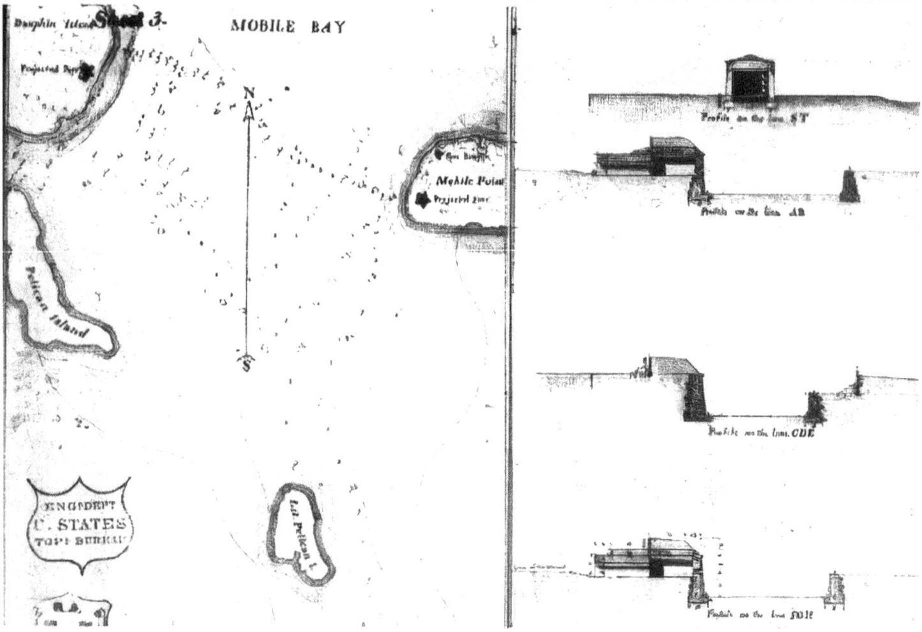

Early plan of the projected twin forts for Mobile Point and Dauphin Island NARA

tain galleries. The service gallery, or "monkey run," nearer the parade had smaller vaulting than the casemate galleries.

Bernard had originally planned for these casemates to fire on the sea and land fronts, but as the design evolved, a higher earthen glacis and covered way made it possible to fire only in defense of the ditch.

Around the entire perimeter of the fort was a high coverface with a cournterscarp to mark the outside of the ditch. So high is the coverface that very little of the fort is visible from the outside. This was to protect the brick scarp from bombardment.

In the original plan, as many as five subterranean brick-vaulted passageways were to have been added to give access between the covered way or the ditch to the outside. These could be used for sorties against besiegers. The defenders would sally forth from them. Today, only two are apparent. One is the long passageway that leads to the main sallyport and the other is on the southeast side. The long passageway that leads to the sallyport originally had double doors on each end. The southeast side passageway required a ladder to enter from the covered way.

Most of the firepower of the fort came from guns en barbette on the second tier. Most of the fire on the main channel would be from two faces, the southwest and west. In order to add more firepower, a lunette gun battery was built around the 1822 Mobile Point Lighthouse. Another battery was added to the west covered way. The lighthouse battery also contained a hot-shot furnace.

Because the citadel was so poorly ventilated its accommodations were uncomfortable in the subtropical heat of southern Alabama. A large three-story officers' quarters with a basement was eventually built outside the fort on the south side.

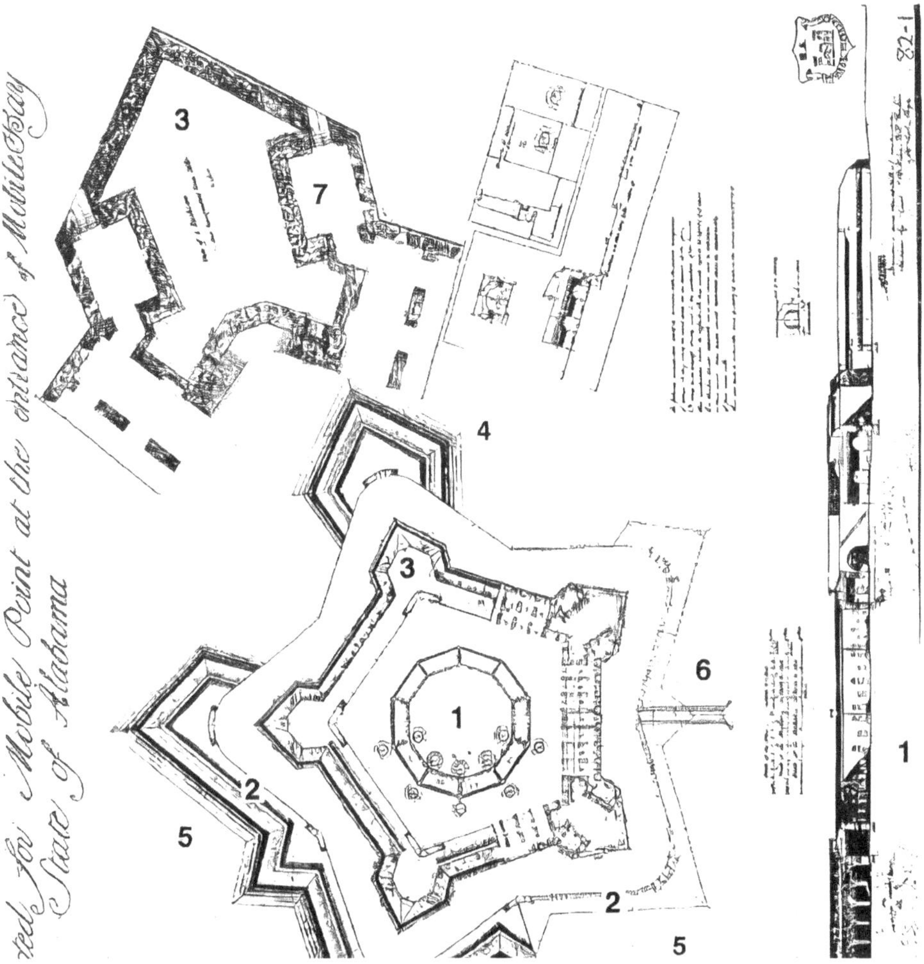

Detailed plan of Fort Morgan. Legend: 1. Citadel; 2. Covered way; 3. Bastion; 4. Lighthouse battery; 5. Glacis; 6. Postern; 7. Bastion flank casemate
Dale Manuel Collection

Two magazines were built into the salients on each side of the main sallyport and were accessible from the parade and galleries. The magazines were vented and wood-lined.

The main sallyport was aligned with the long passageway and the main citadel gate. Two sets of heavy wooden, iron studded, double doors barred entry. In the foyer between the doors are four musket loopholes on each side that could be used to fire on any enemy that had gained entry. Loopholes in the inner door columns and the doors themselves could also be utilized to fire on the foyer. A smaller postern is in the flank of the south bastion.

On January 4, 1861, seven days before Alabama seceded from the Union, Governor Moore ordered the state militia to seize Fort Morgan and Fort Gaines. The Confederates would strengthen the defense of the fort by adding sandbag traverses to the fort's terreplein, add a beach battery, and mount a total of 46 guns of which 11 were

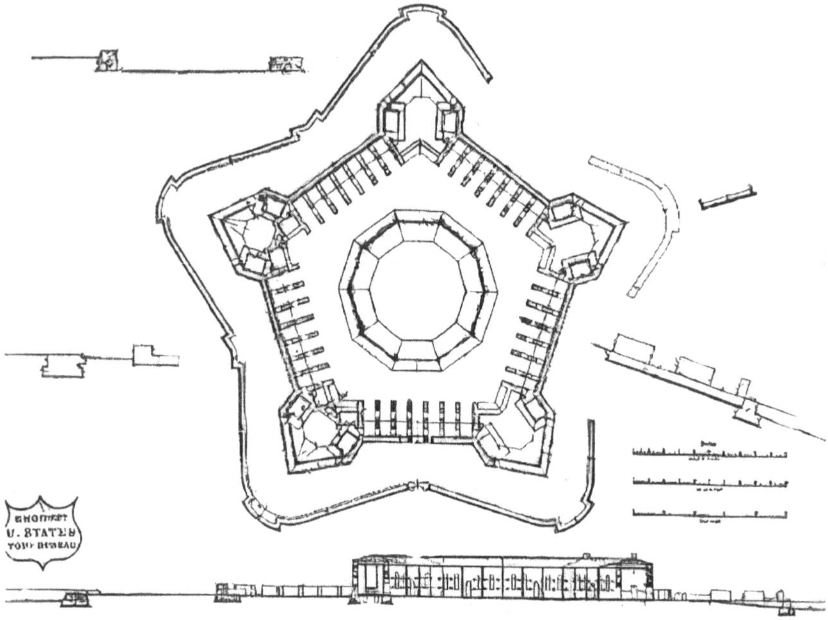

Plan of the fort that shows casemates and bastion flanking casemates
Dale Manuel Collection

Civil War views of the fort's damaged interior and citadel
John Weaver Collection

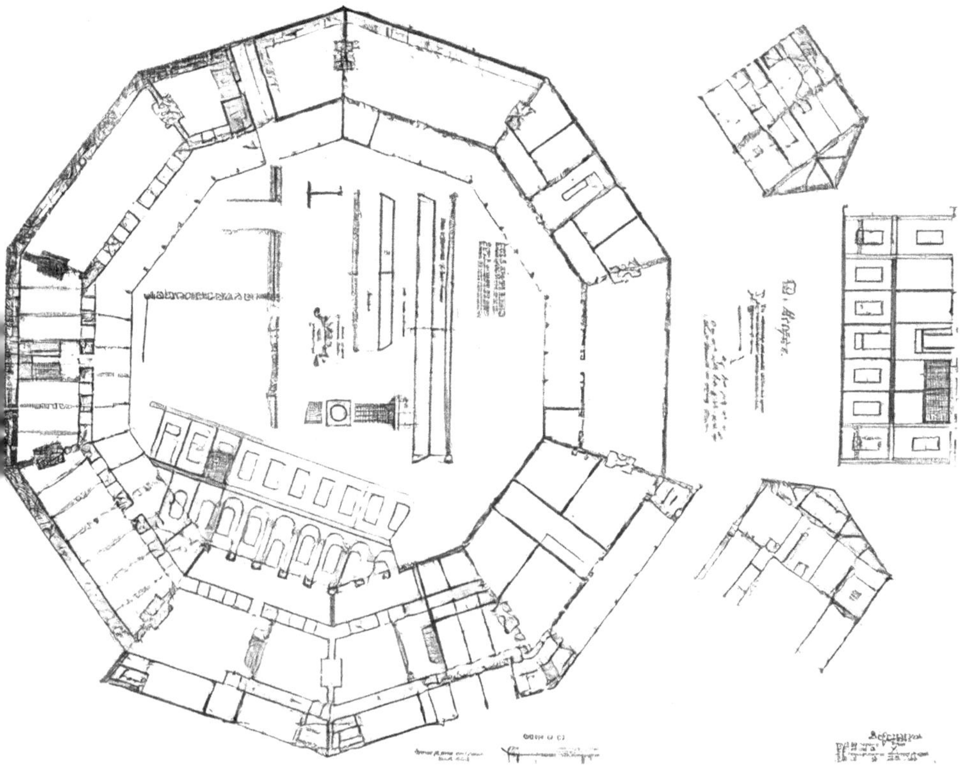

Plans of the ten-sided citadel that was gutted by fire during the Civil War. Only the outline of the inner wall remains Dale Manuel Collection

The Officers' quarters were added outside the fort's outworks. It contained three stories with a basement. The two stair towers had lightning rods on top Dale Manuel Collection

These steps are said to still bear the blood stains of Colonel Stewart. He and three gun crewmen were killed on April 30, 1863 when a rifled 32-pounder gun that was being test-fired burst John Weaver Collection

rifled and 35 were smoothbore. On the morning of August 5, 1864, the fort's true test came as Admiral Farragut's fleet began to run the gauntlet between the torpedo field on the western side of the main channel and the guns of the fort. In command of the fort was a Virginian and long time naval officer, General Richard Page. During the passage of the fleet, 491 projectiles were fired from the fort. On the morning of the 9th, Union troops landed at Navy Cove to start their investment of Fort Morgan. The fort's small garrison labored night and day to build traverses to protect the guns and casemates. The buildings around the fort, such as the hospital, quarters, and stables were set on fire to clear the field of fire. On the 9th, two monitors, three sloops of war, and some gunboats engaged the fort for several hours. The fort was bombarded by the fleet both night and day. The land forces completed their first line of approach across the peninsula on the 10th and the second and third parallels on the 14th to within 700 yards of the fort. The Union sappers continued their work until the 21st when they reached within 200 yards of the ditch. On the 15th, three 15-inch shells struck the right flank of bastion number four and breeched the wall, disabling a flank howitzer. By this time heavy siege guns on the land approach opened fire with a vigorous bombardment. On August 22nd the fleet circled the fort. Guns from land and sea opened up on the fort's beleaguered garrison. This bombardment of 3,000 projectiles disabled all but two of its guns. With the guns of the water and lighthouse batteries unserviceable, they were ordered to be spiked. That night shells caused the citadel to catch fire and burn out of control. The magazines containing eighty thousand pounds of powder were in great danger of explosion. The powder was brought out and flooded. General Page determined he had no means left for defense and on the morning of August 23, at 6 o'clock ordered the white flag to be raised. At 2 o'clock P.M. the surrender was complete. After surrendering, the officers were sent to Fort Warren in Boston Harbor and the men were sent to the prisoner of war camp at Elmira, New York, where as many as half died of smallpox.

In 1874, 12 emplacements were modified on the west and southwest terrepleins in order to mount 200-pounder Parrott rifles. These were the only guns to be mounted in the fort until the Spanish-American War. In 1875, plans were made to construct new exterior batteries on the west outworks of the fort, but work was suspended in 1876. A single service magazine remains of this work that is now part of Battery Thomas.

The 12 Parrott rifles would remain in place for 20 years.

Fort Morgan from the air. Endicott battery has been removed using Photoshop
Photo by John Weaver

Fort Morgan, Endicott Period

When troops arrived at Fort Morgan just prior to the Spanish-American War in 1898, the conditions at the fort were primitive. Battery I of the 1st US Artillery was the first permanent garrison in thirty years. During the Endicott period the fort would increase in size. Six concrete batteries would be constructed, and the fort would become a channel mine depot.

Construction of Battery Bowyer began in 1896 and was completed in 1898. Battery Bowyer was the only Endicott battery at Fort Morgan to be completed in time for the Spanish-American War. The battery mounted four 8-inch rifles on disappearing carriages.

The lighthouse battery once held the brick 1822 and the iron 1872 Mobile Point Lighthouses on this base Photo by John Weaver

One of the curtain angles where a bastion meets a curtain. Each angle contained two flank embrasures. Photo by John Weaver

The battle-damaged Mobile Point Lighthouse John Weaver Collection

Photo of Fort Morgan Endicott Battery Duportail Mark Berhow Collection

During World War I, the guns were removed for conversion to railroad artillery in France. In the early 1920s the left flank of the battery was used as the post switchboard room. A radio station was built on the #3 gun platform.

In 1898, Battery Thomas was completed. The battery was named for Captain Evan Thomas, who was killed in 1873 during the Modoc Indian War at Lava Beds, California. The battery mounted two 4.7-inch Armstrong rapid-fire guns. The battery was built to defend the minefield and channel entrance against fast torpedo boats and minesweepers. By WW I the guns were considered obsolete and were removed. In the early 1920s the #1 gun emplacement at Battery Thomas became the Coincidence Range Finder (CFR) station for Battery Schenck.

Battery Schenck from the mound that makes up Battery Thomas Mark Berhow Collection

Battery Thomas from atop the old fort Mark Berhow Collection

In 1899, Battery Schenck was constructed to guard the channel entrance. The battery originally mounted two 3-inch rapid-fire rifles on masking parapet mounts. In 1903, an additional 3-inch rifle was emplaced on a pedestal mount. The battery was named in honor of 1st Lieutenant William T. Schenck who was killed on June 30, 1898, in action near Subic Bay in the Philippines. In 1920, the Army no longer considered the battery vital for defense and removed two guns. In 1923, the remaining gun was removed when the army abandoned Fort Morgan.

Construction of Battery Duportail began in 1899. Built across the parade of the old fort, the battery held two powerful 12-inch rifles on disappearing carriages that were mounted in 1902. The battery was named in honor of Major General Louis L. Duportail who was Chief Engineer of the Continental Army from 1777 to 1783. The massive concrete battery once contained magazines, two Taylor-Raymond hoists, hoist shafts, and later, power generators. The area of the old fort between the front of the concrete parapet of the battery and the old brick casemates was filled with earth and rubble. The old ditch in front of the battery was also filled. The fill has since been removed from both areas. The brick and earth ramparts of the old fort on the southern side were removed to clear a field of fire for the 12-inch guns.

The east end of Battery Bowyer Mark Berhow Collection

The two 12-inch rifles were Model 1888 and measured over 36 feet in length, with a weight of 52 tons. They could fire a 1,046 pound projectile a distance of 8 ½ miles with a powder charge of 268 pounds. With a muzzle velocity of 2,250 feet per second, a projectile could penetrate 13 inches of armor plate at about three miles.

These rifles were mounted on Model 1896 Buffington-Crozier carriages, better known as disappearing carriages. Two massive swinging gun arms supported the gun. The release of a counterweight into a pit below the carriage raised the gun into the firing position. The recoil from firing returned the gun downward and to the rear into

Gun crew is firing a large steel rifle on a "disappearing carriage."
Dale Manuel Collection

The post buildings from the Mobile Bay side of the reservation
Dale Manuel Collection

View of the post from the Northeast bastion of the old fort
Dale Manuel Collection

the loading position. This action concealed the gun and provided safety for the gun crew. Battery Duportail was deactivated in 1923. In 1942, the guns were scrapped.

Battery Dearborn was completed in March of 1900. The battery was named in honor of a hero of the War of 1812, Major General Henry Dearborn. The battery

Photo of the post canteen Dale Manuel Collection

In this 1907 photo, the repaired west Battery Commander's (BC) station is shown. The station is shown next to a storm-wrecked instrument column. Dale Manuel Collection

mounted eight 12-inch Model 1890 mortars with four mortars in each of the battery's two pits.

The Experimental or Test Battery was completed in 1916 to test the effects of naval gun fire on a 14-inch Taft era battery. An old 10-inch rifle on a disappearing carriage was mounted in the emplacement. Small arms target silhouettes were placed in the battery and goats were tethered in the magazine, plotting room, and galleries to determine the effects of detonation on humans. In March, 1916, 55 shots were fired at the battery in three phases by two ships. The USS New York with 14-inch guns and the USS Arkansas with 12-inch guns fired from ranges of 5800, 8500, and 16,000 yards with twelve hits on the target. There was no permanent damage done to the gun and carriage. Few of the silhouettes were damaged and none of the goats were injured.

Starting in about 1900, mining facilities were constructed outside the rear of the old fort. A mine casemate was built into the glacis of the old fort. A narrow-gauge tramway linked the cable tank, torpedo storage, and torpedo loading room to the engineer's wharf that doubled as a mine wharf. Next to the torpedo loading room was a small dynamite service room.

An old view of the post from Battery Dearborn Dale Manuel Collection

The two-pit mortar battery, Battery Dearborn Mark Berhow Collection

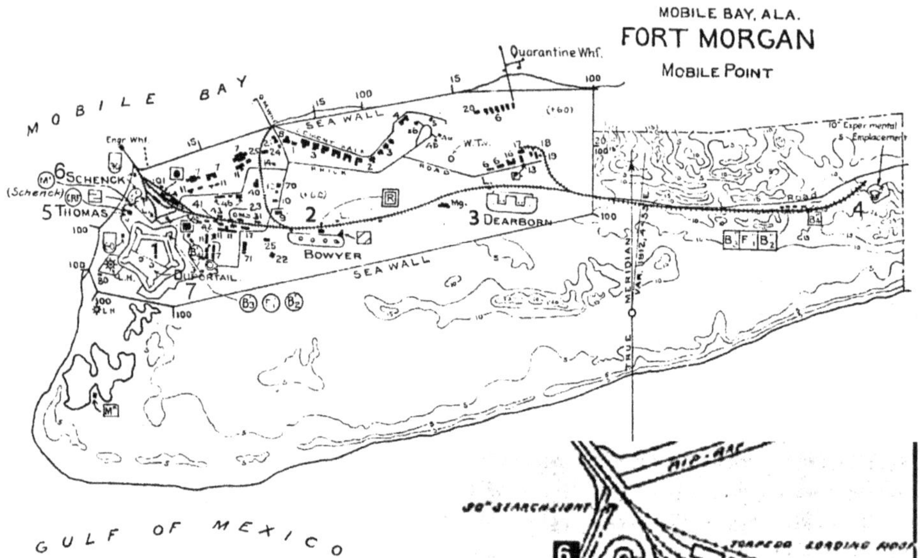

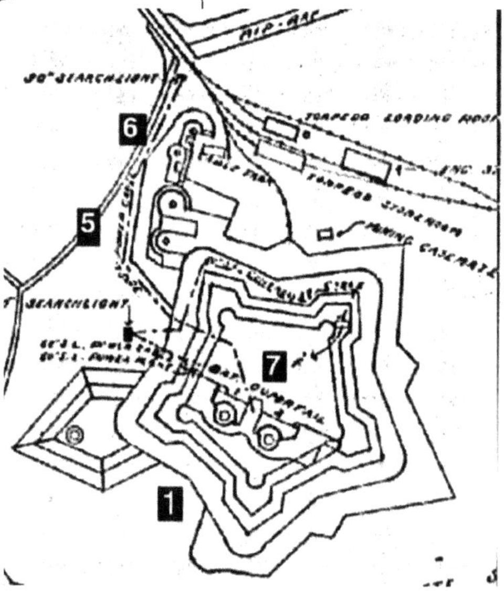

Fort Morgan Legend: 1. Old Fort; 2. Battery. Bowyer; 3. Battery Dearborn; 4. Experimental Battery; 5. Battery. Thomas; 6. Battery Schenck; 7. Battery Duportail Dale Manuel Collection

The post at Fort Morgan contained almost one hundred buildings that had been completed by 1910. The buildings included: commanding officer's quarters, commissioned officers' quarters, non-commissioned officers' quarters, barracks, administration building, hospital, morgue, guardhouse, post exchange, gymnasium, mess halls and lavatories, bakery, electric power and ice plant, fire station, firemen's quarters, coal shed, stable, wagon shed, telegraph station, quartermaster's wharf, quartermaster's office, oil house, carpenters' shops, post office, artillery engineer's office, ordinance storage house, engineer's office, and blacksmith shop. Each of the four enlisted men's barracks could house one hundred men. Seven 30- and 60-inch searchlight stations covered the channel and Gulf beach waters at night with projectors.

In 1902, the "Peace Magazine" was built between Battery Dearborn and Battery Bowyer for drier storage of the fort's powder during times of peace. In the event of war, the powder would be moved to the better-protected magazines of the batteries. The copper-roofed magazine was used until the fort was abandoned in 1924.

Battery Duportail from the back and of the gun platforms
Mark Berhow Collection

To aim or point the battery's guns, systems of fire control and position-finding were used. Fire control dealt with the identification of targets and distribution of fire while position-finding determined the range and direction of a moving target. In a horizontal base system, primary and secondary base stations were used to establish a base line for a triangular plot. By taking simultaneous sightings at timed intervals from these two locations, the target's range and direction (azimuth) could be calculated with the aid of a plotting board.

On September 26, 1906, a disastrous hurricane hit Fort Morgan. Every building on the post was either damaged or destroyed. The nearby village of Navy Cove was swept away. After the storm, work began on a concrete seawall that would encircle the post.

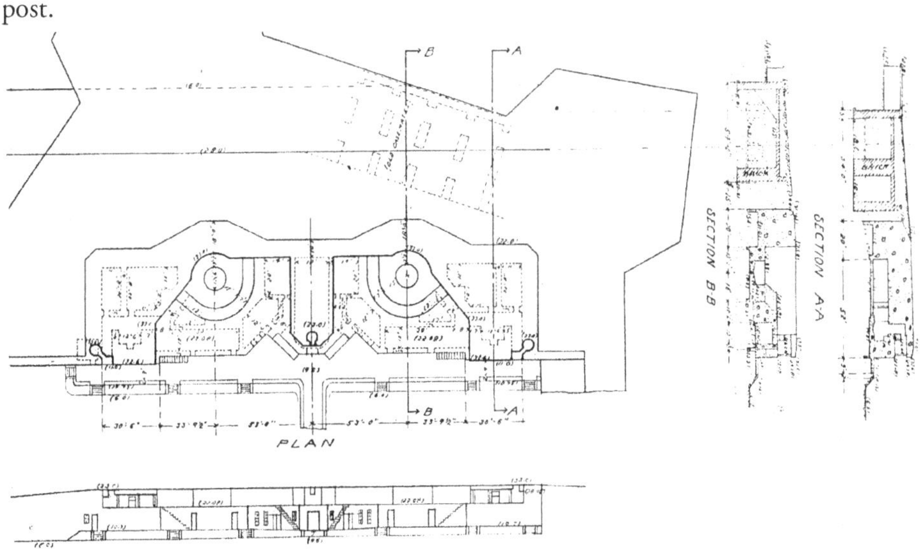

Battery Duportail once mounted two 12-inch rifles on disappearing carriages
NARA

12-inch projectiles. Dale Manuel Collection

Fort Morgan in the World Wars

During WW I, Fort Morgan served as an artillery training center. Training included coast artillery, anti-aircraft, and trench-mortar warfare. Many of these troops would serve on the battlefields of France. The fort's garrison would grow from 400 to more than 2,000 men. Temporary barracks and other buildings would have to be built to serve the tremendous growth. After the war the fort would be closed in 1923. For the first time since 1898 the fort was without a military presence.

In 1927, the State of Alabama purchased the old fort reservation to create a state park. Labor by the Public Works Administration during the 1930s created the state park facilities.

In late 1941, the fort was reactivated and was manned by units of the Army, Coast Guard, and the Navy. The Navy stationed the minesweeper U.S.S. *Guide* to search for any mines that may have been laid by German submarines. Coast Guardsmen patrolled the beaches on former U.S. Cavalry horses that had been transferred from the Army.

In April, 1942, Battery "F' and other elements of the 50th Coast Artillery Battalion served the 155mm GPF batteries that were positioned on circular concrete

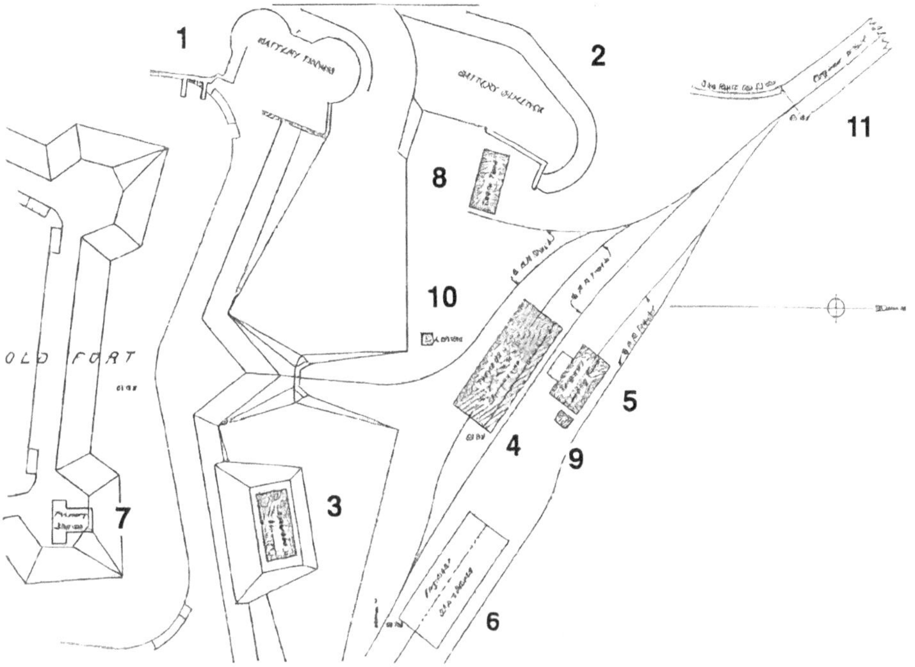

Plan of the channel mining facilities. Legend: 1. Battery Thomas; 2. Battery Schenck; 3. Mining Casemate; 4. Torpedo Storage; 5. Torpedo Loading Room; 6. Engineer Storehouse; 7. Primary Station; 8. Cable Tank; 9. Dynamite Service Room; 10. Latrine; 11. Engineers' Wharf Dale Manuel Collection

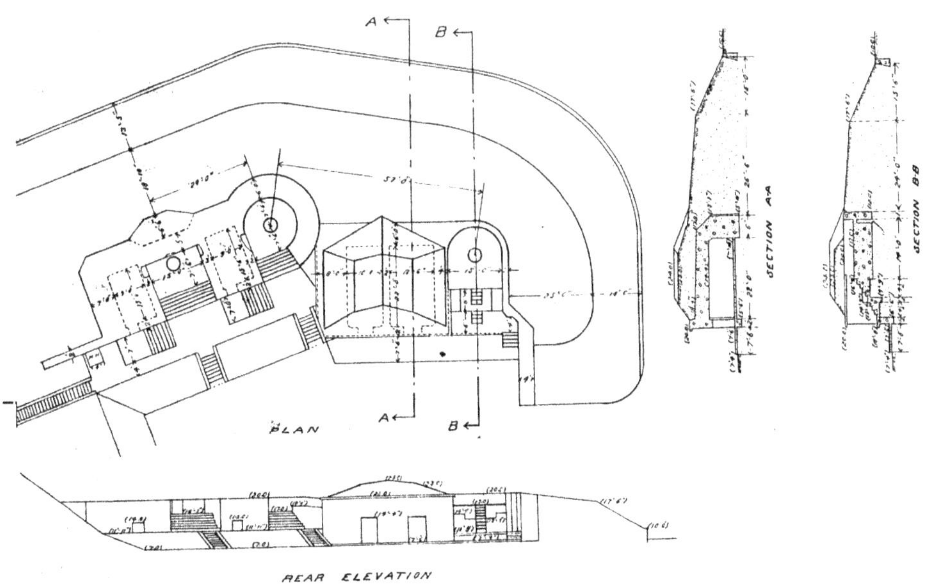

Battery Schenck once mounted three 3-inch rapid-fire rifles NARA

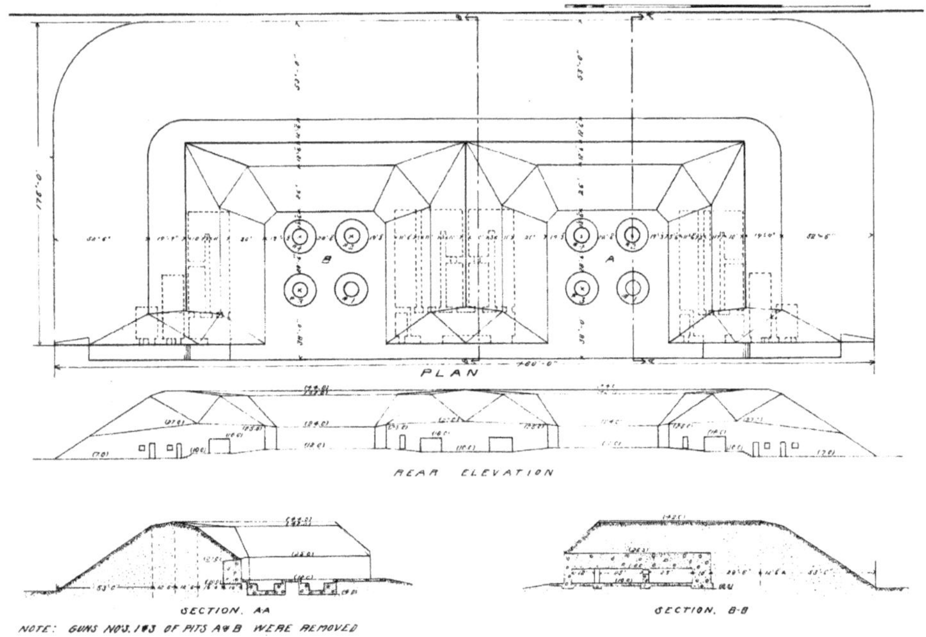

Battery Dearborn was a two-pit mortar battery that once mounted eight 12-inch heavy mortars. It had shell and powder magazines, data booths, crow's nests, and later generator rooms NARA

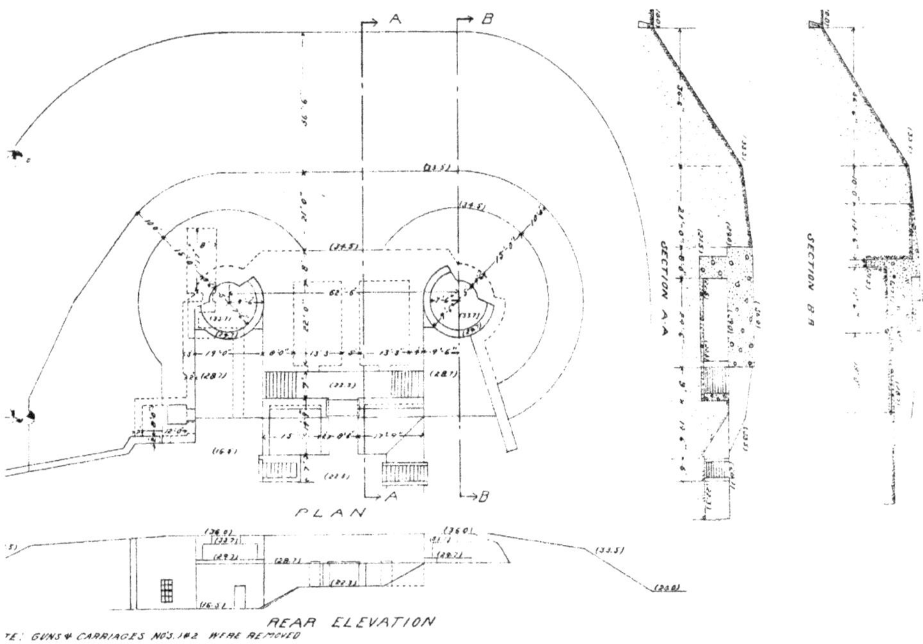

Battery Thomas once mounted two 4.7-inch Armstrong rifles NARA

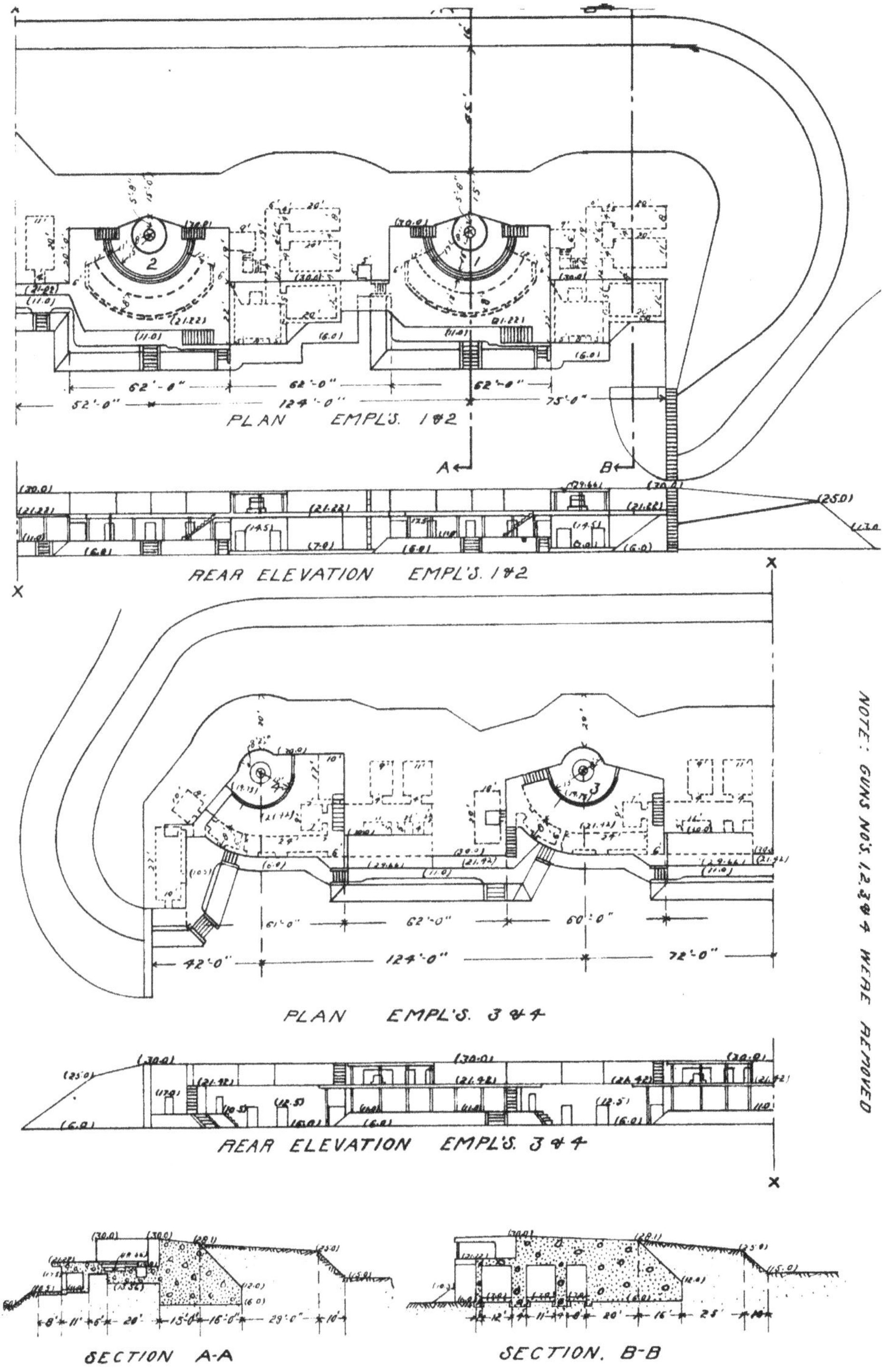

Battery Bowyer once mounted four 8-inch rifles on disappearing carriages NARA

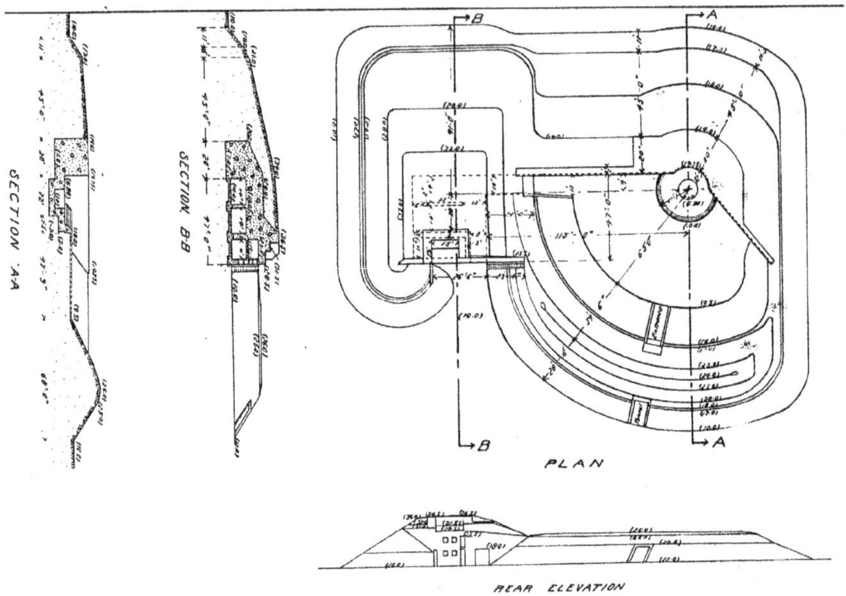

The Test Battery once mounted an old 10-inch rifle on a disappearing carriage
NARA

The remains of the engineer wharf that was also used for mining
Dale Manuel Collection

Today, there is not much left of officers' row. Time and storms have taken their toll on the old frame buildings at the fort. Many were lost during the hurricane of 1906. Mark Berhow Collection

mounts, "Panama Mounts", on top of two of the old fort's bastions. The guns were Model 1918 Mls on Model 1918 carriages. One Panama Mount still remains on the southwest bastion of the old fort and one of the guns is displayed in front of the fort.

The US Model 1918Ml, 155mm gun was copied from a French-designed artillery piece for the Army. During the 1920s and 1930s the Army adapted these guns for coast defense. The weapon could be cheaply and quickly deployed on Panama Mounts. The gun could fire a 95-pound projectile a maximum range of 11 miles. The fort was closed after WW II and eventually returned to park status.

1908 view of the multi Primary Base Stations on top of the northeast bastion Dale Manuel Collection

The mostly-buried Test Battery Mark Berhow Collection

Model 1918 M1, 155mm gun that is on display outside the old fort
Mark Berhow Collection

One Panama Mount is still on top of bastion #4 Dale Manuel Collection

Fort Gaines, Dauphin Island Begun 1819 and 1854

The current bastioned pentagonal fort, built on the eastern tip of Dauphin Island, is actually the second Fort Gaines. The original Fort Gaines was begun in 1819 and was to have been the twin of Fort Morgan across the entrance of Mobile Bay on Mobile Point. The original Fort Gaines, like Fort Morgan, was designed by Simon Bernard to be one of the first works of the Third System of coastal defense. If the original Fort Gaines had been completed, it would have been a massive bastioned regular pentagon with a large decagonal citadel in the center of the parade.

In 1821, a controversy arose that caused work on the original Fort Gaines to be discontinued. Some members of Congress believed that the channel covered by Fort Gaines was shallow and could not be navigated by sailing ships of war with deep-draft keels. Since only a fraction of the work on the citadel had been constructed at a cost to that point of $219,743.91, some believed the cost of the project was too great. Others argued that the original fort would be necessary to keep an enemy from using the island as a base for invasion of the mainland as it had been used by the British in their

Gorge area of the fort Photo by John Weaver

campaign against Fort Bowyer in the War of 1812. The fort lay unfinished for decades. Only a small portion of the decagonal citadel had been constructed.

In time, war steamers with shallower drafts that could navigate the Dauphin Island Channel were developed. Steamers using that approach would be out of the range of the guns at Fort Morgan. Funds were appropriated in 1845, but title disputes delayed work for about nine years. On January 20, 1853, final title of the property was acquired by the United States Government. In 1854, the current Fort Gaines was begun. The remains of the old citadel were leveled to make room for the new fort. One

Curtain and tower bastion. A 32-pounder is on top. Photo by John Weaver
Sallyport with display of Parrot rifles Photo by John Weaver

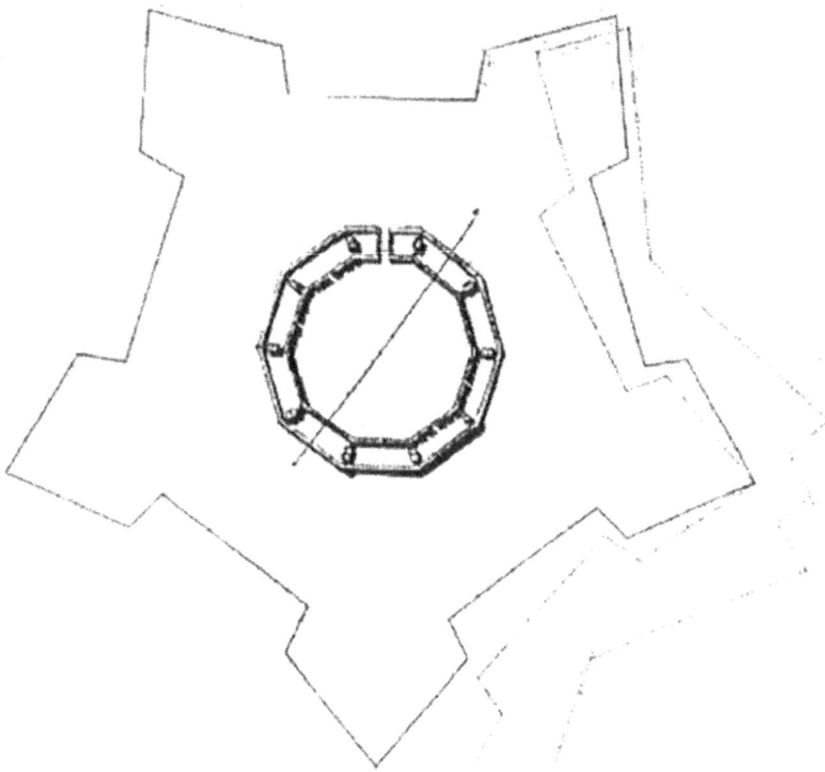

Only a small portion of the older fort's citadel had been constructed
Dale Manuel Collection

Inside a tower bastion looking toward the magazine Photo by John Weaver

plan of the new fort shows the outline of the old citadel in the northwest corner of the present parade. The present fort was designed by Joseph Totten, one of the finest military engineers this nation has ever produced. The design is more pragmatic, less costly, and smaller in scale than the grandiose Fort Morgan. Most of the construction was complete by 1861. The Confederate States of America would make the finishing touches to the fort. Danville Leadbetter came to Mobile in 1853 to supervise construction of Fort Gaines and repairs to Fort Morgan. He would later design many of Mobile's defenses and become a Confederate general. P. G.T. Beauregard was also associated with the construction of Fort Gaines as the chief supervisor of the Gulf Coast forts. He, too, would become a Confederate general.

Sea-face curtain with musket loopholes and tower bastion
Photo by John Weaver

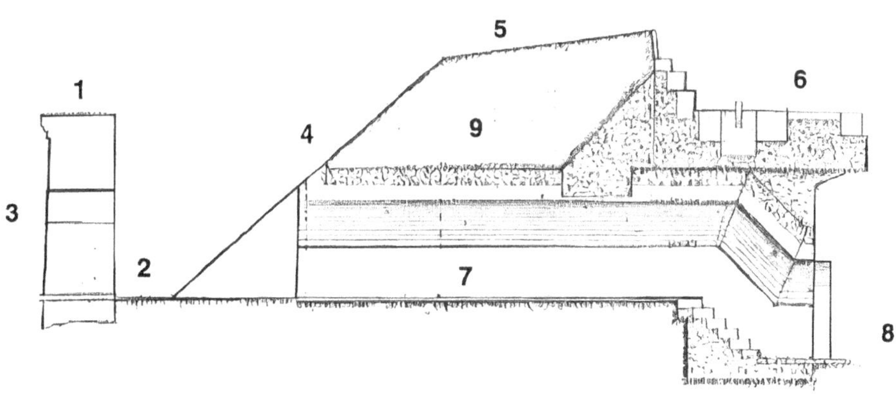

A detached scarp: 1. Carnot Wall; 2. Chemin de Ronde; 3. Loophole; 4. Inferior Slope; 5. Superior Slope; 6. Terreplein; 7. Gallery; 8. Parade; 9. Rampart
Dale Manuel Collection

South parapet and gun emplacements Photo by John Weaver

In 1853, the fort was named in honor of General Edmund Pendleton Gaines (1777-1849). As commandant of Fort Stoddert, he arrested Aaron Burr in 1807. In the War of 1812, as a colonel, his regiment was the rear guard in the American retreat from the battle of Chrysler's Field. As a general he successfully defended Fort Erie in 1814. He was a veteran of the Black Hawk War of 1831. He was wounded in the Second Seminole War of 1835. During the Texas Revolution in 1836, he held Nacogdoches for five months. During the Mexican War, Gaines recruited volunteers from Southern states. He died in New Orleans on June 6, 1849.

Fort Gaines is an irregular pentagon with five tower bastions. The design is very similar to that of Fort Clinch, Florida. The design made use of two technological advances. One is the tower bastion. The other development, originated in France by Napoleon's Minister of War, Lazare Carnot, was a detached scarp with musket loopholes. Behind the scarp was a narrow passage called a chemin de ronde where riflemen

Even the latrine had musket loopholes to fire on the slopes
Dale Manuel Collection

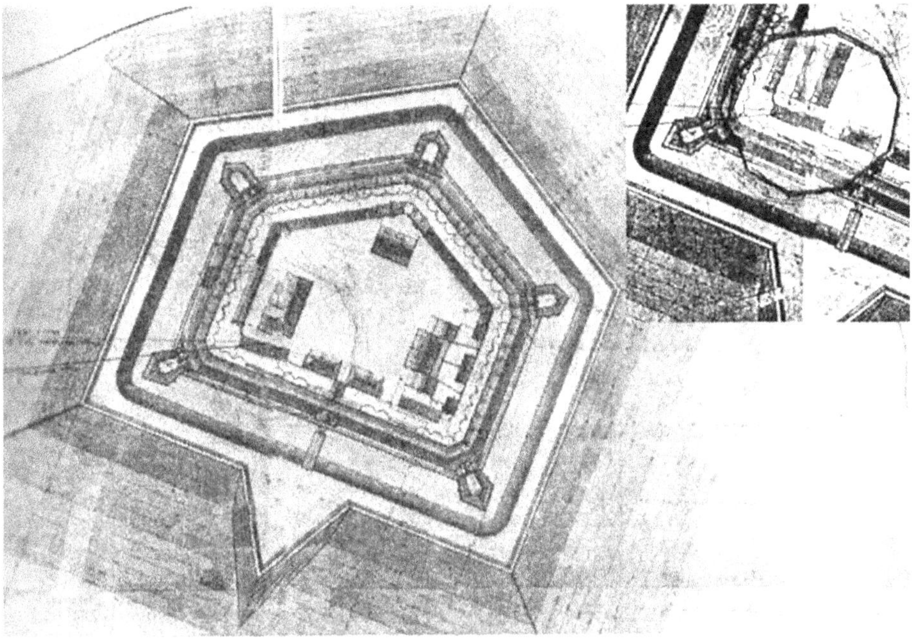

Plan of Fort Gaines shows the location of the old incomplete citadel with a detailed inset Dale Manuel Collection

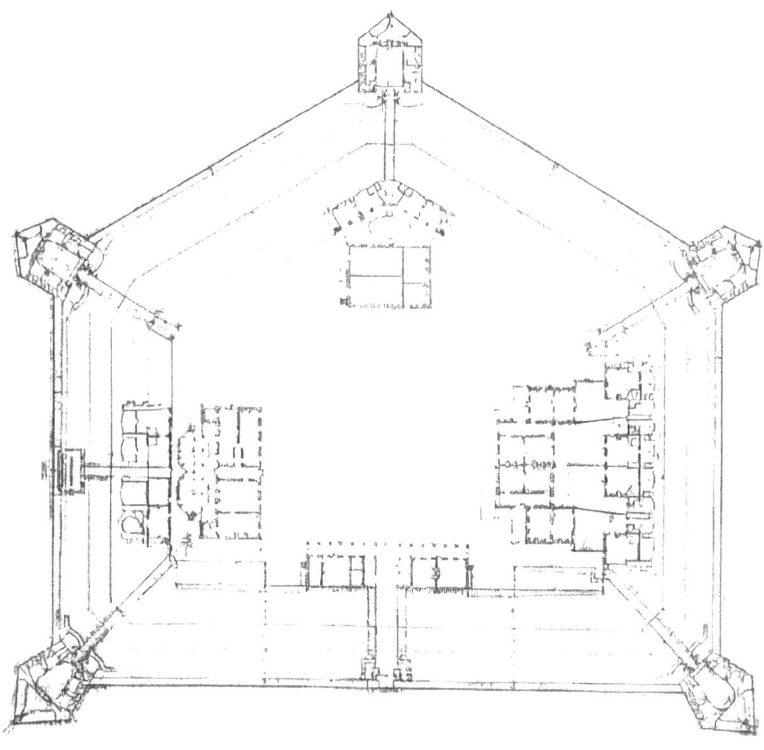

Details of the fort's construction Dale Manuel Collection

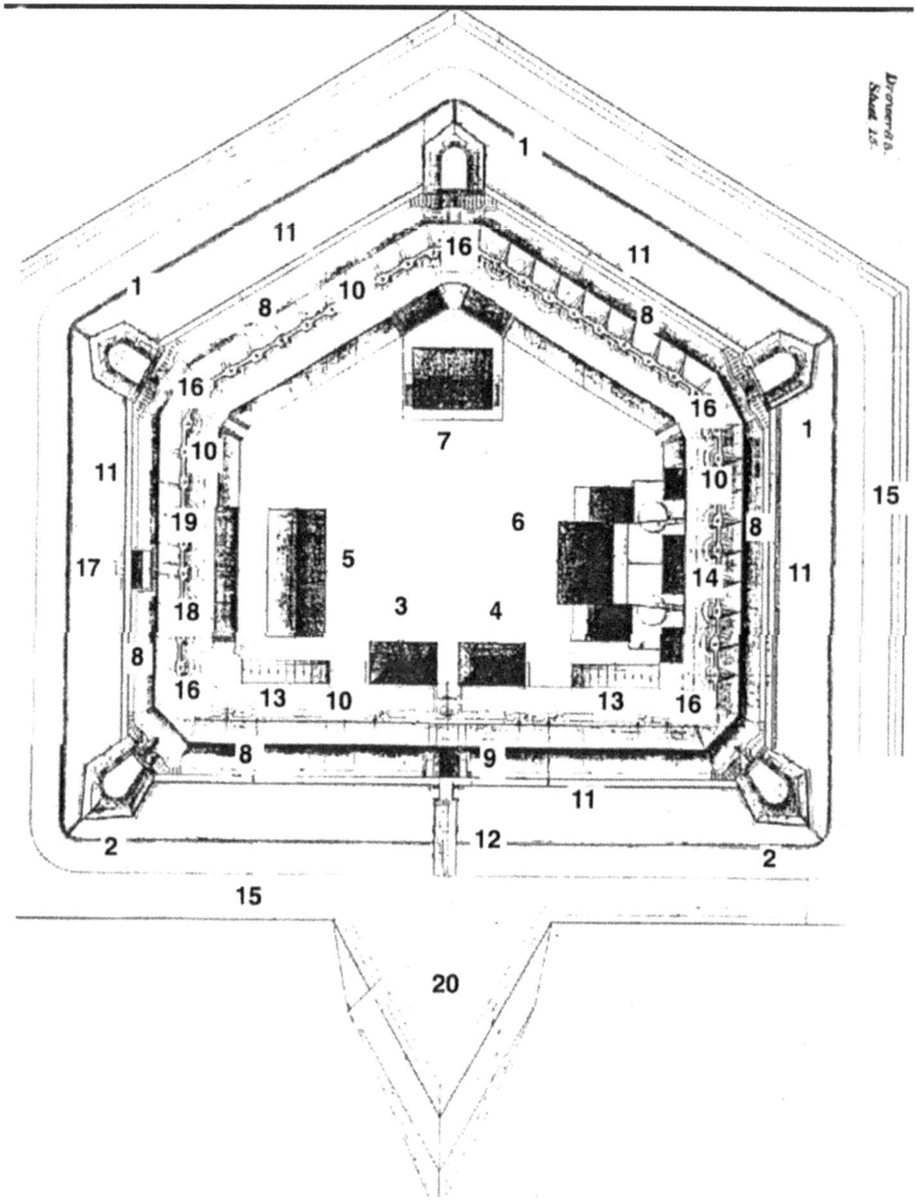

Details of Fort Gaines. Legend: 1. Channel Bastions; 2. Gorge Bastions; 3. Guardhouse; 4. Headquarters; 5. Barracks that were not build according to these plans, but were built outside the wall;. 6. Officers' quarters; 7. Quarter Master's office and stores; 8. Chemin de Ronde; 9. Sallyport; 10. Terreplein; 11. Carnot Wall; 12. Drawbridge; 13. Ramps; 14. Kitchens; 15. Glacis; 16. Galleries; 17. Latrine; 18. Bakery; 19. Blacksmith Shop; 20. Re-entering place d'armes
Dale Manuel Collection

From this pre-Civil War drawing of Fort Gaines some buildings can be identified 1. Two-story Q.M. office & storehouse; 2. Three-story officers' quarters; 3. Headquarters; 4. Guard House; 5. The outside two-story barracks. Dale Manuel Collection

could fire through the loopholes. Starting at this narrow passage, the exterior slope of the earthen rampart rose to the superior slope and finally to the parapet that had a masonry revetment. Behind, the parapet, guns were mounted en barbette. The idea was that even if the scarp of the Carnot wall was battered down or breached, the earthen rampart, and the guns atop it, would remain intact.

The tower bastions are higher than the curtain. Each tower bastion has two tiers. The first tier's barrel-and-groin vaulted rooms contained four flank howitzers and four flank musket embrasures. The musket loopholes were narrower, but higher than the howitzer loopholes so muskets could be aimed toward the top of the gently sloping glacis that surrounded the work. Each tower contained a magazine. To store rainwater, cisterns were below the floor of each bastion. A rainwater collection system channeled water collected from the top tier through pipes to the cisterns. A spiral staircase led up to the barbette tier where a gun was mounted. Each tower could sweep the faces on both sides with flanking fire. The two gorge tower bastions have different flanking angles of the bastion than the three channel towers. Each gorge tower has an 80-degree flanking angle of the bastion while the channel bastions have 105 degrees of traverse.

The fort's only entrance, called a sallyport, is in the middle of the gorge wall. It once had a drawbridge that spanned a dry ditch. When the drawbridge was raised the entrance to the sally port would be blocked. Great doors in the sallyport also barred entry. The sallyport opened up to a large parade.

A parapet with gun emplacements ran around the entire perimeter as did the chemin de ronde. Each tower bastion was reached through a covered gallery or tunnel that ran under the ramparts. The gallery also gave entry to the chemin de ronde. On both ends of the gorge are gun ramps that lead up to the terrepleins of the ramparts.

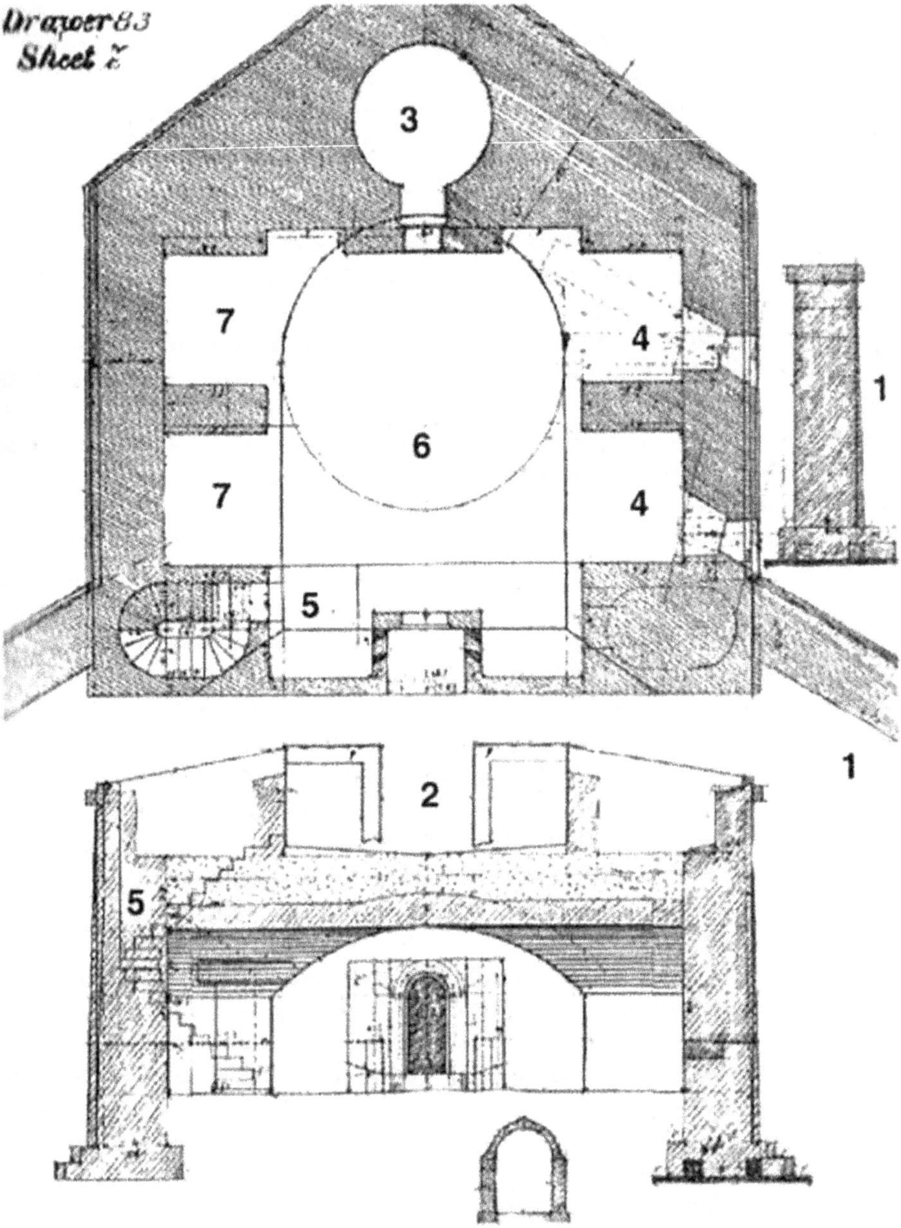

Plan of a channel tower bastion. Legend: 1. Carnot Wall; 2. Barbette gun emplacement; 3. Magazine; 4. Flank gun embrasures; 5. Staircase; 6. Vaulted Casemate; 7. Flank gun emplacements Dale Manuel Collection

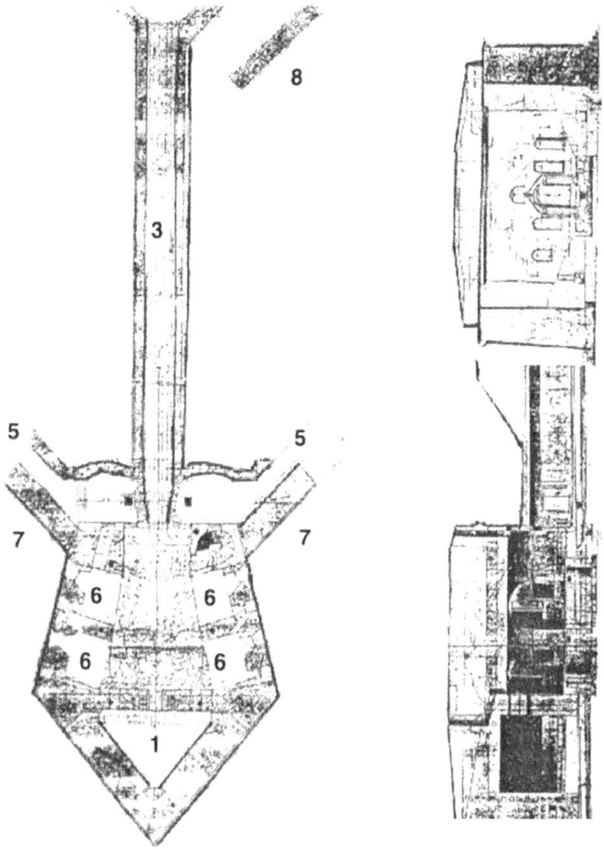

Legend: 1. Magazine; 2. Barbette gun emplacement; 3. Gallery; 4. Parade; 5. Chemin de Ronde; 6. Flank gun emplacement; 7. Carnot Wall; 8. Parapet
Dale Manuel Collection

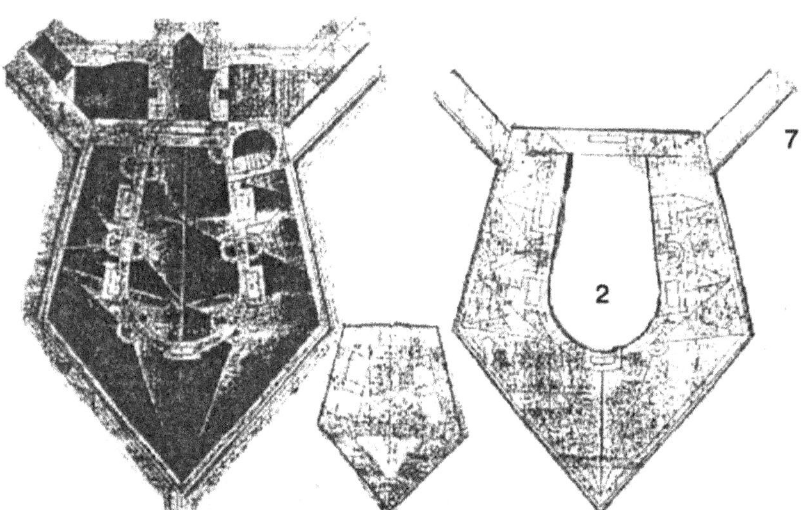

Plan of one of the gorge tower bastions Dale Manuel Collection

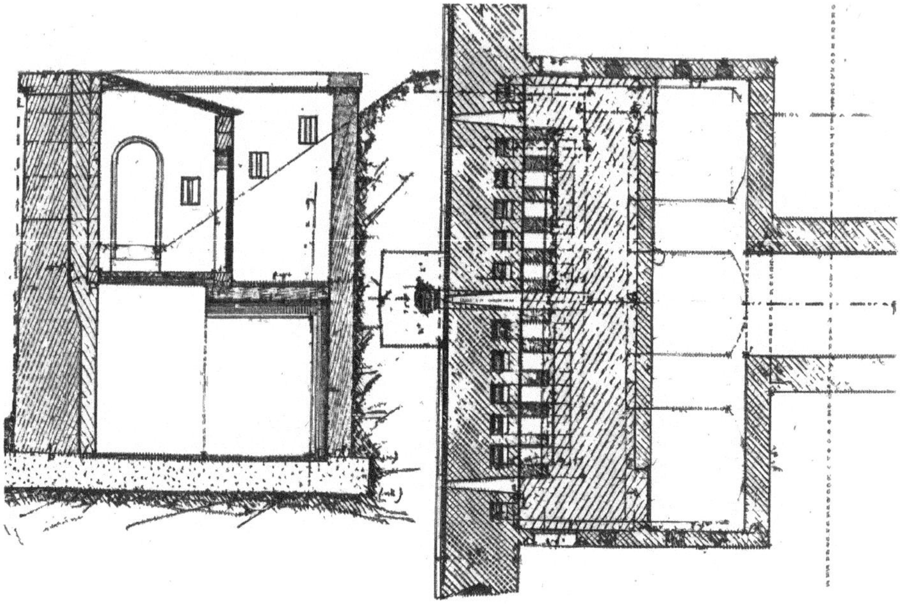

Plan and section of the three-story officers' quarters. During the battle, Union shells damaged the two top stories. Dale Manuel Collection

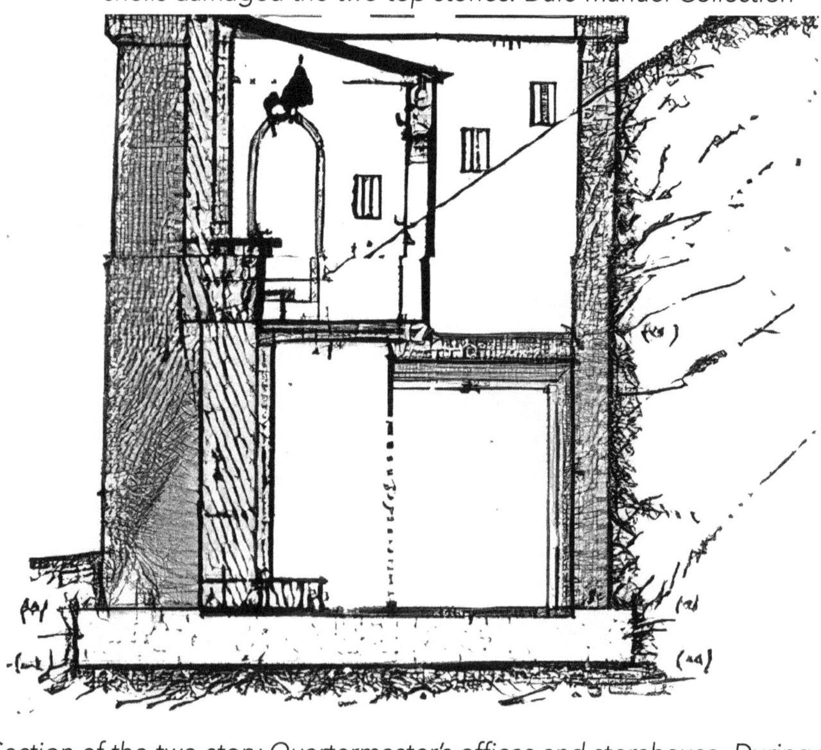

Section of the two-story Quartermaster's offices and storehouse. During the battle its top story was damaged. Part of the building was removed to make room for Battery Stanton during the Endicott period. Dale Manuel Collection

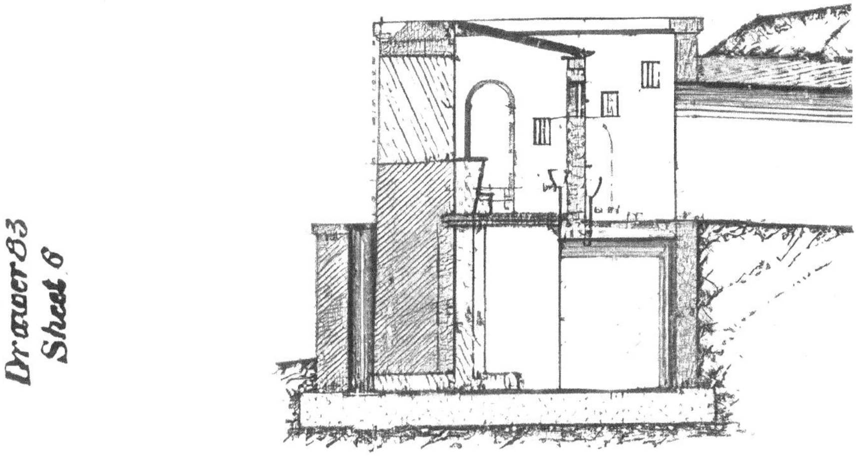

Plan of the latrine. A gallery led from the parade to the latrine. Outer doors connected the chemin de ronde passage.

A number of buildings are on or line the parade. On the right side of the sallyport was the guardhouse, now used as the museum's gift shop. On the left side of the sallyport was the Commandant's office, now used as restrooms. On the north side of the parade is a bakery and a blacksmith shop. In the middle of the bakery and blacksmith shop is another gallery that leads to a latrine on the north curtain. Passages on both sides of the latrine lead to the chemin de ronde. Hot-shot furnaces were on both sides

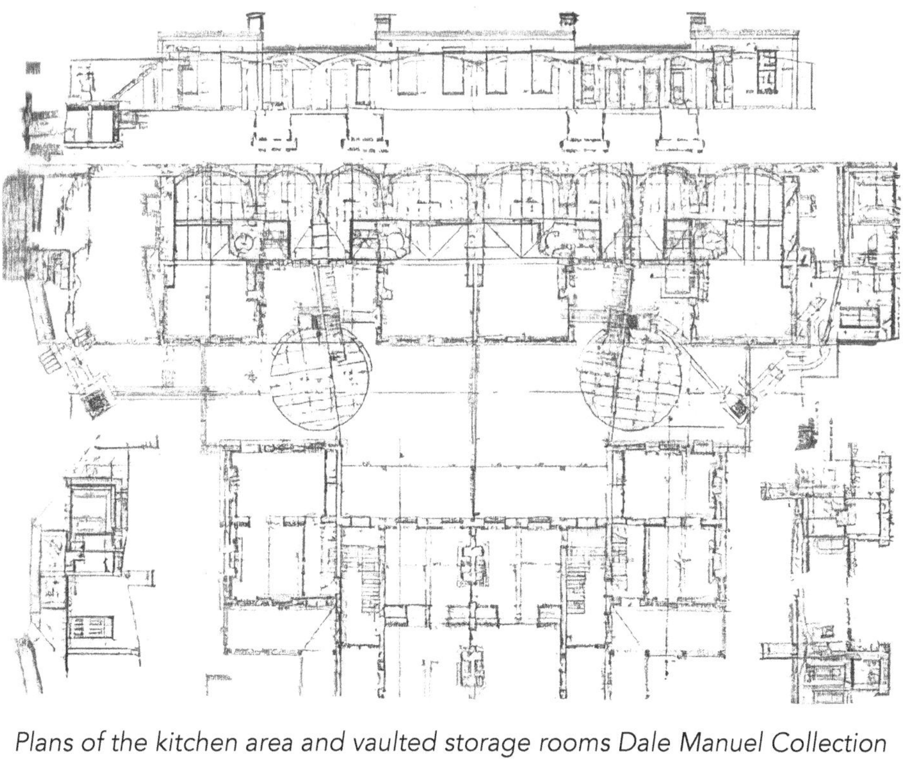

Plans of the kitchen area and vaulted storage rooms Dale Manuel Collection

The interior of a kitchen Dale Manuel Collection

of the channel faces. In the center of the parade's east salient were the Quartermaster's offices and storerooms. On the south side of the parade are the remains of the officers' quarters, once three stories high. In back of the officer's quarters were two cisterns fed by the rain runoff from the roof of the quarters. A vaulted kitchen was built under the terreplein of the south wall. Vaulted store rooms were built there as well.

Early plans show that a two-story barracks was to have been built in the northwest corner of the parade for the garrison. This barracks was not built at that location. Instead, a barracks was built outside the fort's walls near the sallyport.

On August 4, 1864, 1,500 Union troops landed on the western end of Dauphin Island. As the Federal troops began moving toward Fort Gaines they met Confederate skirmishers. The next day the Union troops slowly advanced on the fort and eventually formed siege lines. At the same time Farragut's fleet was forcing its way into the bay. The Confederate skirmishers were ordered back to their entrenchments around the fort. Reinforcements from Mobile had arrived and included boys from the Pelham Cadets. The garrison had grown to about 800 men. The barracks outside the fort had to be burned to clear the field of fire for the fort's guns.

After the naval battle was over, the monitor USS Chickasaw and other ships bombarded both Fort Powell and Fort Gaines. For three days Fort Gaines was bombarded from land and sea. On August 7, Union troops began to bring up rifled guns and heavy mortars to reduce the fort. The fort's commander, Colonel Anderson, was concerned for the safety of his men. With most of his guns being unserviceable he could not match the artillery barrage the Federals could hurl at the fort. On the evening

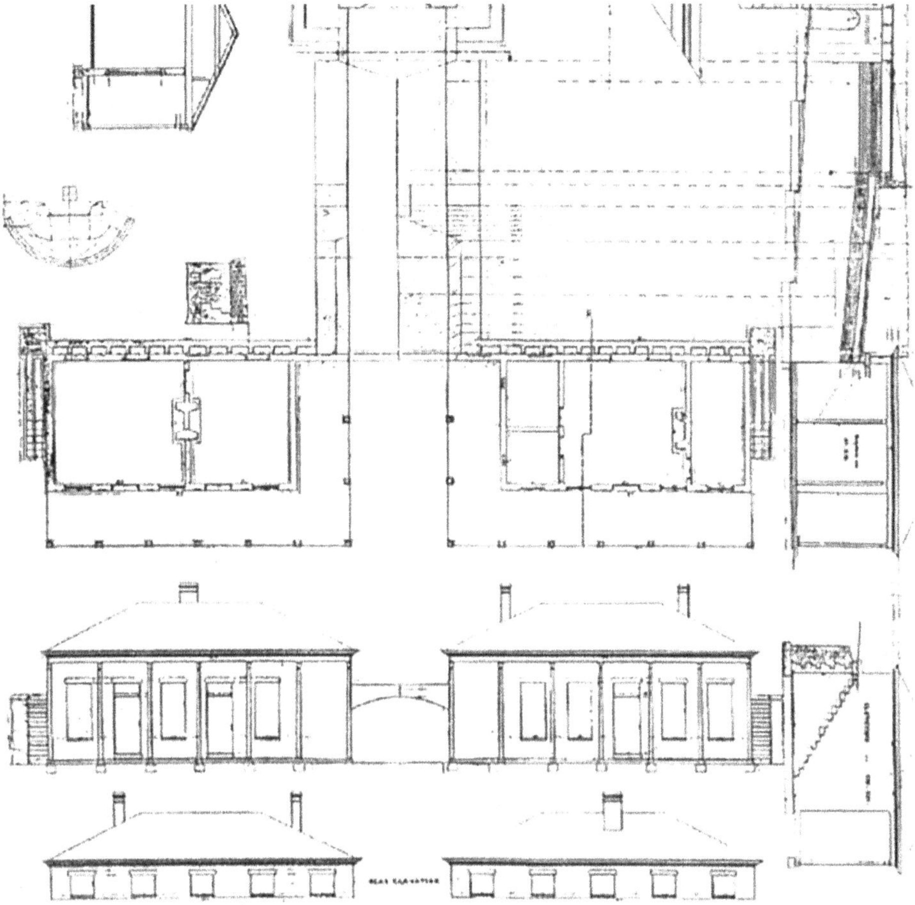

Plans of the headquarters and guardhouse with the sallyport in the center
Dale Manuel Collection

of August 7, Colonel Anderson rowed to Farragut's flagship, *Hartford,* to discuss the terms of surrender of the fort. Despite the insistence of General Page, the commander of Fort Morgan, not to surrender, Fort Gaines was surrendered the next morning, August 8. The battle of Dauphin Island was over. The prisoners were transported to the prisoner-of-war camp on Ship Island, Mississippi.

After the Civil War, Parrott rifles were mounted on circular barbettes. The fort would not be re-armed until the Spanish-American War. The fort fell into disrepair after it was closed in 1923. In November, 1955, the Dauphin Island Park and Beach Board opened the fort as a museum after cleaning and restoration. It is open year round with a small admission fee.

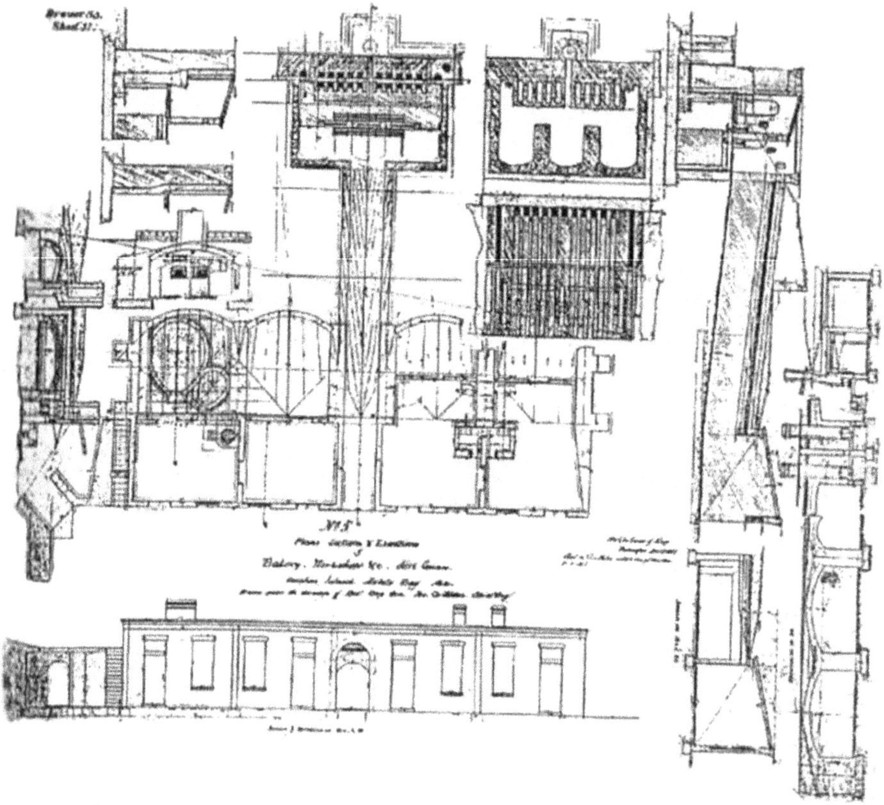

Plans and elevation of the bakery, blacksmith shop, and latrine Dale Manuel Collection

The bakery contained two ovens, the one on the left being larger than the right one Dale Manuel Collection

The forge in the blacksmith's shop Dale Manuel Collection

The officers' quarters were once three stories high Mark Berhow Collection

This building contains a bakery, a blacksmith shop, and the latrine gallery
Photo by John Weaver

Left was the headquarters; right was the guardhouse and the center sallyport
Photo by John Weaver.

Remains of the Quartermaster's office at Fort Gaines Dale Manuel Collection

The remains of the Quarter Master's office and storehouse at Fort Gaines would have looked identical to the Quarter Master's office and storehouse at Fort Clinch, Florida. Many of Fort Clinch's buildings were identical to that of its sister Fort Gaines. Since Fort Clinch was never bombarded by Union troops and never had an Endicott battery on its parade, most of its buildings are intact.

Fort Gaines from the water Mark Berhow Collection

After-battle view of Fort Gaines drawn by Captain John Gordon Dale Manuel Collection

The 9,000-pound anchor from Admiral Farragut's flagship USS Hartford. The old wooden sail-steamer was preserved by the Navy for many years but sank in 1956 at her mooring in the Norfolk Naval Yard. Dale Manuel Collection

The parade of the moderate-sized bastioned pentagon Fort Gaines Photo by John Weaver

Fort Gaines from the air. Endicott battery removed using Photoshop to show original design Photo by John Weaver

Landward front of Fort Gaines Photo by John Weaver

The door in the sallyport leads to the Chemin de Ronde Dale Manuel Collection

Fort Gaines, Endicott Period

Starting in 1898, a double 8-inch breech-loading rifle battery was constructed in the interior of the old fort, behind the southeast channel face. The rifles were mounted on modified 15-inch Rodman carriages. This early battery was later covered by the construction of Battery Stanton's number one emplacement.

Begun in 1901, Battery Stanton would have three 6-inch rifles on disappearing carriages. Battery Stanton was named in honor of Captain Henry W. Stanton, 1st U.S. Dragoons, who was killed in action by Apache Indians in the Sacramento Mountains, New Mexico, on January 19, 1855. Battery Terrett was constructed outside the north side of the old fort. Originally Battery Terrett had only two 3-inch rapid fire rifles. Later another 3-inch rifle on a masking parapet mount would be added. Battery Terrett was named in honor of 1st Lieutenant John C. Terrett, who was killed at the battle of Monterey, Mexico, on September 21, 1846.

During the Endicott period, the Fort Gaines Reservation would cover most of the east end of the island. Other facilities on the reservation included officers' quarters, non-commissioned officers' quarters, barracks, mess hall, artesian well, Q.M. storehouse, U.S. engineer's building, a CRF station, and a Q.M.'s wharf. The original area of the reservation was 983.9 acres, but it was reduced to about 267 acres in 1911.

A concrete Coincidence Range Finder (CRF) station for Battery Stanton was constructed on top of the southwest tower bastion of the old fort. In order to aim the guns

Endicott Battery at Fort Gaines Mark Berhow Collection

Battery Stanton from the fort's parade Mark Berhow Collection

#2 gun platform of Battery Stanton. This emplacement once mounted a 6-inch rifle on a "disappearing carriage." In all, Battery Stanton would mount three of these rifles. Mark Berhow Collection

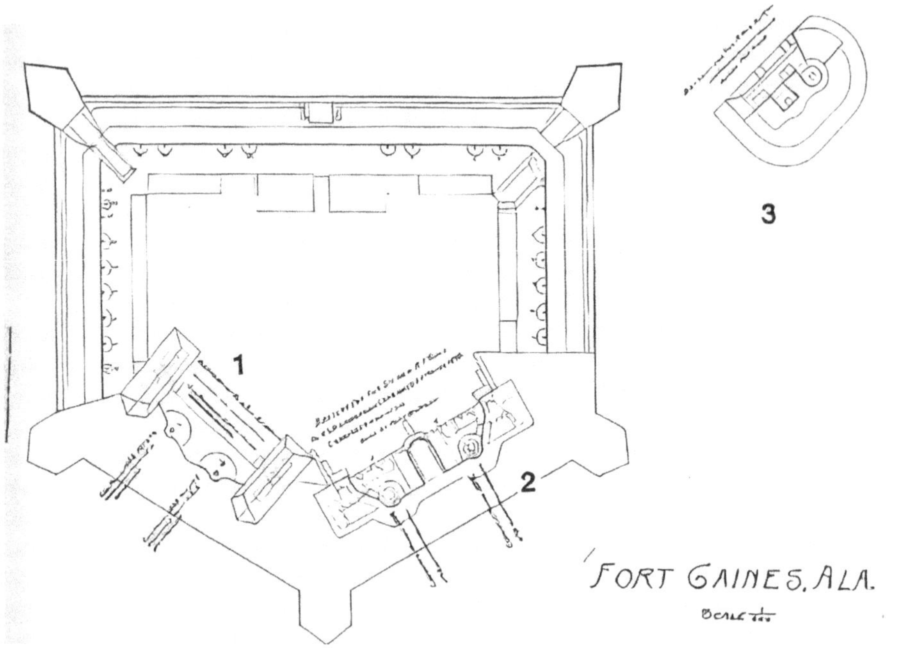

Plan shows the early Endicott batteries at Fort Gaines. Legend: 1. Two 8-inch BL rifles on Rodman carriages; 2. Battery Stanton; 3. Battery Terrett
Dale Manuel Collection

#1 emplacement of Battery Stanton that once held a 6-inch rifle on a disappearing carriage. The emplacement was built on top of the two 8-inch Breech-Loading rifle batteries. Mark Berhow Collection

of the battery with pinpoint accuracy, this self-contained base station determined the range and direction (azimuth} of the target.

The batteries of Fort Gaines would serve until after WW I. In 1923, Fort Gaines was closed, along with Fort Morgan.

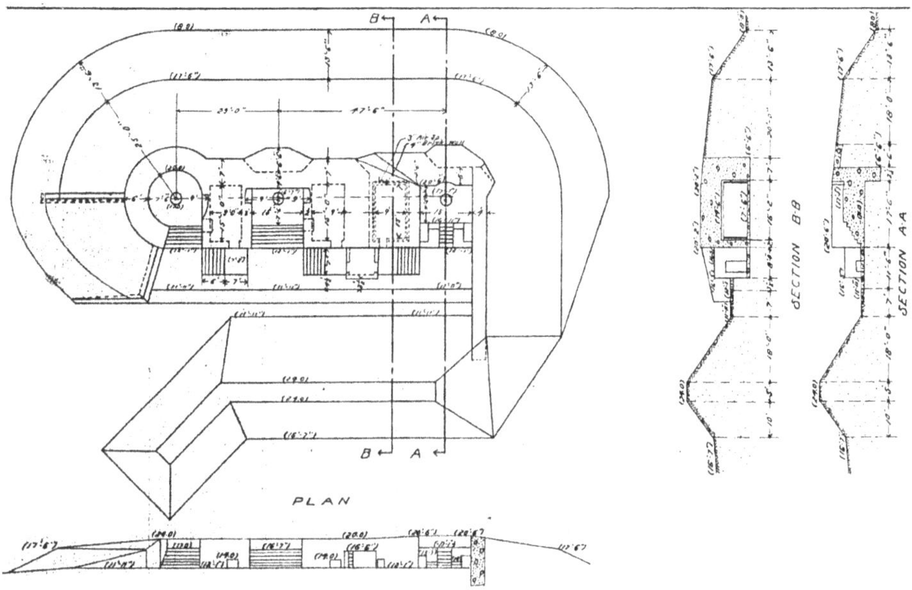

Plan for Battery Terrett which once mounted three 3-inch rapid-fire rifles NARA

Coincidence Range Finding (CRF) station for Battery Stanton was built on top of the southwest tower bastion Photo by John Weaver

Overhead rails in Battery Stanton once delivered ammunition from the magazines to the hoists. Hoists raised ammunition to the gun platform level.
Mark Berhow Collection

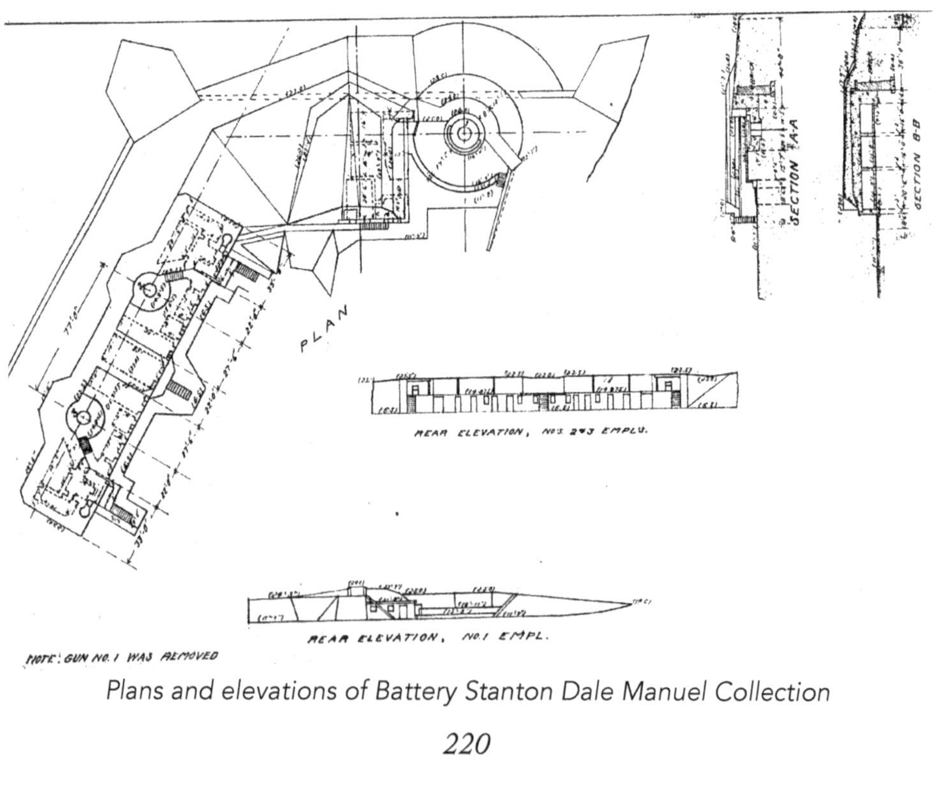

Plans and elevations of Battery Stanton Dale Manuel Collection

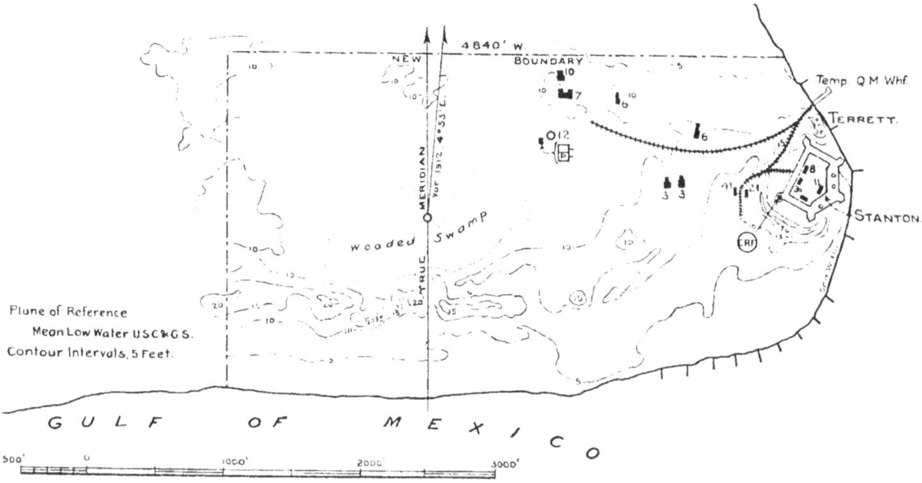

Map of Fort Gaines Reservation during the Endicott period. Legend: 1. Old Fort; 2. Bat. Stanton; 3. Officers' Quarters; 4. Bat. Terrett; 6. N.C. Officers' Quarters; 7. Barracks; 8. Guardhouse; 10. Mess Hall; 11. Original Storehouse; 12. Artesian Well; 21. Q.M. Storehouse; 41 US Engineer's Building NARA

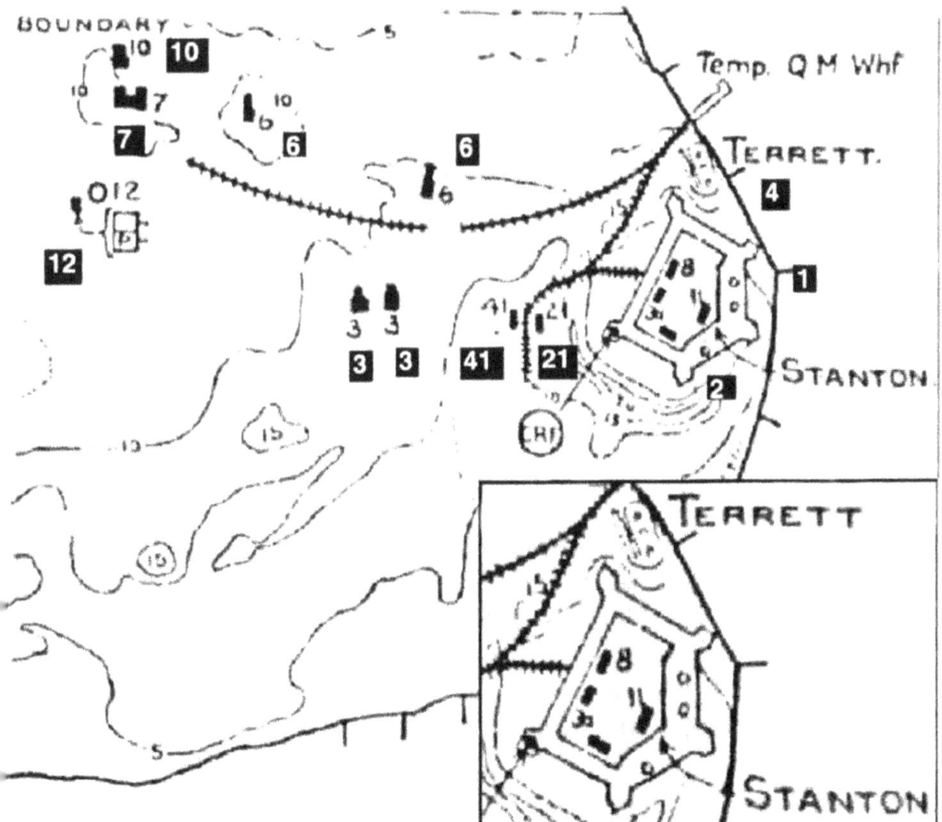

Later plan of the reservation. Note that the wharf was moved to the Gulf side of the island. Dale Manuel Collection

Fort Powell, Grant's Pass Begun During the Civil War

Built on a small oyster bank only 50 feet from the channel of Grant's Pass, Fort Powell was a small earthwork fort that was a marvel of Confederate engineering. It was constructed to guard the back door to Mobile Bay from the Mississippi Sound. In the upcoming battle, the Union Navy would place a high priority on capturing the diminutive fort. The first action of the Mobile Campaign would be the long-range bombardment of the fort.

The fort was named for Colonel William L. Powell who was in command of Fort Morgan and the construction of Fort Powell. Under his command, construction dragged on at a slow pace. The shell bank was enlarged to about half an acre and raised to three feet above high tide by the use of cribs. A row of cribs was placed in the shape of the trace of the fort along the outside line. Each crib was five feet wide, ten feet long and six feet high and was made of pine logs. The cribs were filled with sand. Rocks were placed on the outside and the inside line of cribs was filled with oyster shells. The Mississippi Sound side, or west face, was completed first. The parapet was 8 feet high and 25 feet thick. On each face of the work, three guns were mounted in a conical-shaped chamber that measured 18 feet in diameter at the bottom and 24 feet at the top. The chambers were formed by sandbag traverses. The traverses were 9 to 12 feet higher than the parapet and formed embrasures that allowed a field of fire of 110 to 170 degrees That could be adjusted by adding or removing sandbags. The parapet was 3 feet above the gun platforms. The guns were mounted 60 feet apart with three on each face.

Gun platforms were carefully constructed in each gun chamber. Four layers of sawn timber sills were laid to form an octagonal platform. An iron pintle and pintle-cross was secured to the center of the platform by iron bolts.

A large mound of sand covered the center bombproof. On the cavalier or top center, a parapet was built. Around the cavalier, rifle pits were dug to provide for musket fire against boat attacks. Behind the center parapet two Whitworth guns were mounted. The bombproof contained the main powder magazine, a filling room, a laboratory,

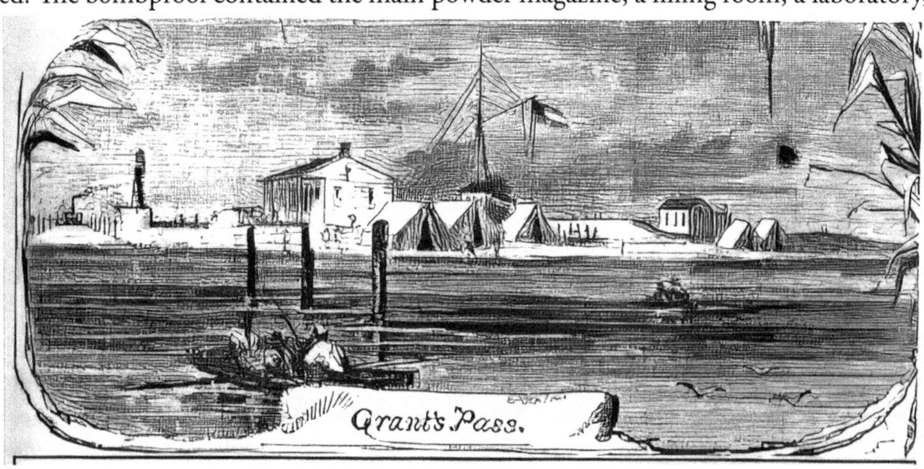

Sketch of Fort Powell

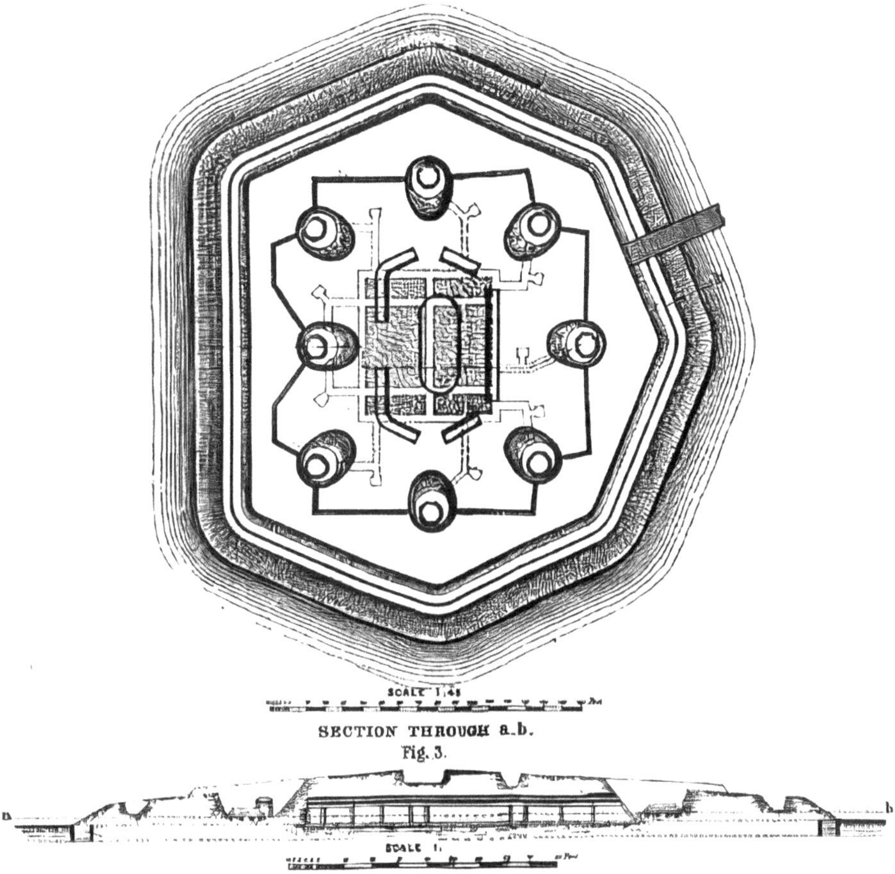

Plan and section of Fort Powell Dale Manuel Collection

This Civil War view of Grant's Pass toll and lighthouse station shows a battery and was likely the location of Fort Powell Dale Manuel Collection

This ten-inch "Confederate Rodman" was rifled to 6.4 inches and was salvaged from the site of Fort Powell. It is now displayed at the Museum of Mobile.
Dale Manuel Collection

a blacksmith shop, water tank, surgeon's room, and quarters for the men. Reinforced tunnels led to each gun and each gun's ready magazine.

The seacoast front was armed with one 8-inch Columbiad, one 6.4-inch rifled "Confederate Rodman", and two 7-inch Brookes rifles. The gorge face mounted a 10-inch Columbiad and a 7-inch Brooke rifle. Since the gorge was incomplete, these guns were exposed from the platform up. The channel west of the fort had been obstructed by torpedoes and a row of chevaux-de-frise constructed of railroad iron.

Fort Powell was commanded by Lieutenant Colonel James M. Williams of the 21st Alabama Regiment. Between February 22, and March 2, 1864, a steady bombardment was kept up on the fort from Union mortar schooners and gunboats on the Mississippi Sound. During this time, not a single gun had been dismounted nor any serious damage done to the traverses. Shell craters were easily covered by throwing sandbags into them. Only one man was killed and another was wounded. Only the wharf and quarters on the east face were damaged by the bombardment.

On August 5, 1864, Federal gunboats shelled the fort at long range from the sound. Admiral Farragut forced his way into the bay and threatened the unfinished and exposed east gorge face. On the same day, the two-turreted monitor USS Chickasaw dueled with Fort Powell. The monitor scored hits on the fort and the fort hit the monitor three times with its 7-inch Brooke rifle. One shell penetrated the fort's bombproof, but did not explode. Knowing that the shell exploding would likely have killed the entire garrison, the decision was made that the fort was untenable and that it must be evacuated. Around 10:00 P.M. that night, the garrison waded to Cedar Point. Colonel Williams, the last man out, lit a slow fuse that blew up the fort's main magazine. The huge explosion happened a half-hour after Colonel Williams had left the fort.

The Naval Battle of Mobile Bay, August 5, 1864

The Confederates had strengthened Mobile's harbor defenses by obstructions and by planting nearly 200 torpedoes on the west side of the bay entrance, forcing all ships to pass close to the guns of Fort Morgan. Although several ironclads were under construction at Mobile, only four Confederate vessels would be engaged in the Battle of Mobile Bay. The steamer gunboats CSS *Selma,* CSS *Gaines*, and CSS *Morgan* were only armored around their boilers. The pride of the small Rebel Flotilla was the flagship of Confederate Admiral Franklin Buchanan, the CSS *Tennessee.* She was a huge ironclad ram that mounted six Brooke rifles and had armor six inches thick. She was 209 feet in length and 48 feet in beam. The casemate that shielded her battery and crew was 78 feet 8 inches long by 8 feet high.

In the upcoming battle the Rebel flotilla would be out-gunned 192 guns to 22 guns. The officers and crews of the Union fleet numbered about 3,000 officers and men, compared to about 470 for the Confederates. The 14 Union ships that would attempt to run the gauntlet consisted of six first-class steam sloops and gunboats, and four monitors, two of which were double-turreted. In order to obtain a low profile in the water, buoyancy in the monitors had to be sacrificed. Lack of buoyancy would prove to be fatal for the USS *Tecumseh.*

Scene from the naval battle just before the surrender of the CSS Tennessee
Dale Manuel Collection

On the morning of August 5, 1864, at 5:30 A.M., the long line of Federal vessels began moving with the tide toward Fort Morgan. Four ironclad monitors moved ahead. Fourteen wooden vessels were lashed together in pairs with the larger vessels on the side that faced the guns of the fort. The first shots at the fort were fired by the monitor USS *Tecumseh* at 6:47 A.M. As firing commenced between the two fleets and the fort, thunderous fire and smoke belched from the guns. The *Tecumseh* struck a torpedo and sank in 25 seconds. One hundred and fourteen officers and men sank to the bottom of the channel. When the USS *Brooklyn* saw the monitor sink it halted and stalled the progress of the fleet. Devastating fire from the Confederates was poured into the Union vessels. At this time Admiral David Farragut climbed the rigging of the Flagship USS *Hartford* to get a better view and ordered the flagship to pass the Brooklyn with the famous words: "Damn the torpedoes! Full speed ahead." Although the passing fleet brushed the torpedoes, amazingly, none exploded, for corrosion and exposure to salt water had taken their toll on them.

As the USS *Hartford* passed the torpedoes, the three Confederate gunboats ran close by, firing their guns. At the same time the CSS *Tennessee* tried to impede the progress of the Union fleet. At 8:40 the fleet was past the guns of the fort. The consorts of the larger ships were unlashed and began chasing the Confederate gunboats. The CSS *Gaines* was hit twice and began to sink. Her crew grounded her near Fort Morgan. The USS Metacomet caught up with the CSS *Selma* just after 9:00 o'clock and a short but fierce battle ended with the surrender of the *Selma*. Most of her officers were either killed or wounded. The CSS *Morgan* escaped but became grounded. That night she would make her escape back to Mobile.

With the Federal fleet anchored four miles up the bay, Admiral Buchanan boldly attacked them in a single vessel, the *Tennessee*. A fierce battle was fought at close quarters as the *Tennessee* was rammed twice by the *Monongahela*, by the *Hartford*, and by the *Lackawanna*. At one point, the *Tennessee* and the *Hartford* brushed alongside each other. The *Tennessee* only fired one shot due to defective primers while the *Hartford* fired a full broadside with little effect on the ram. The *Hartford* was nearly sunk when

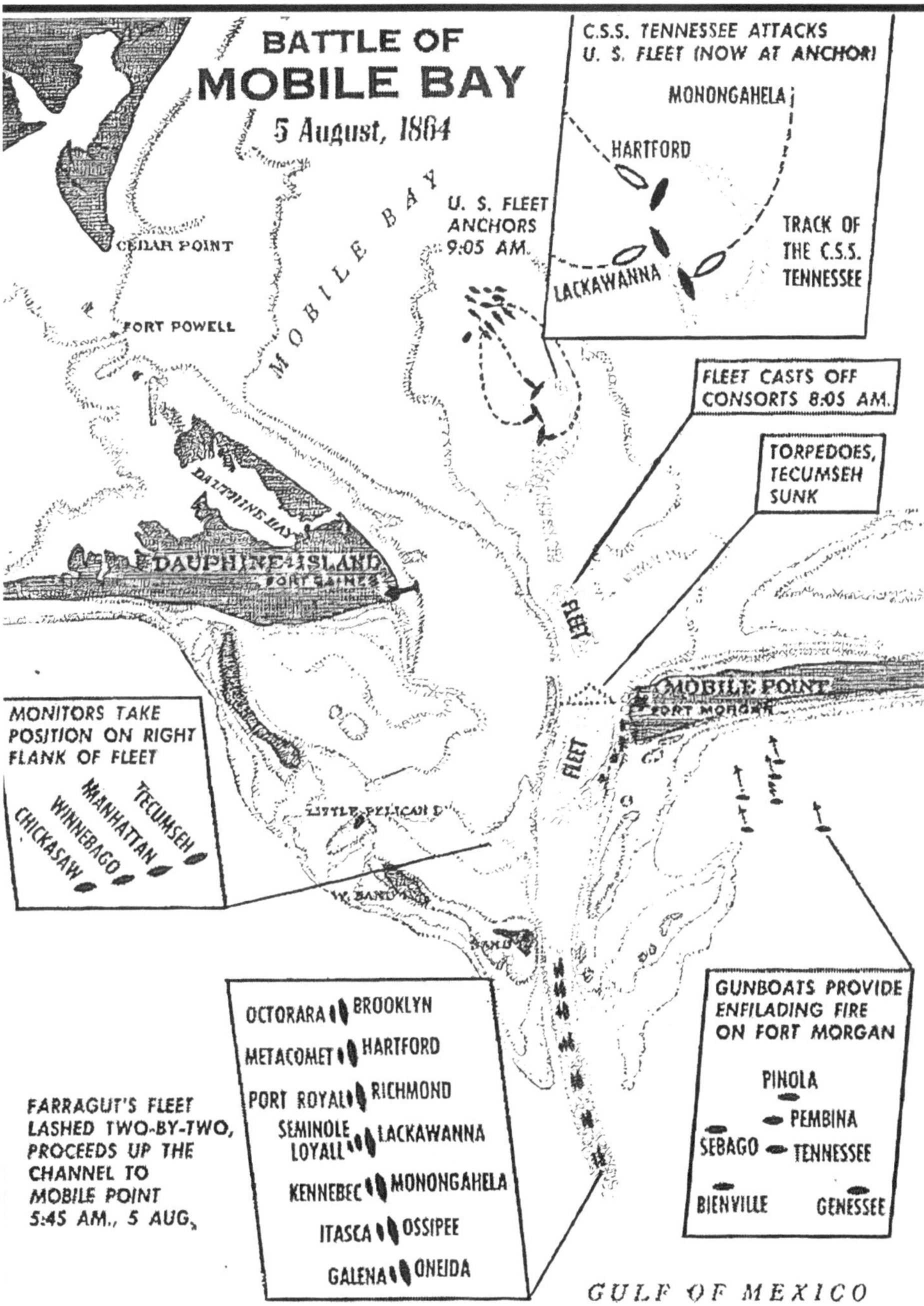

Map of the Battle of Mobile Bay Dale Manuel Collection

Photo # NH 60335 USS Tennessee (ex-CSS Tennessee), circa 1865

The huge ironclad ram Tennessee seen after her capture at Mobile Bay
Dale Manuel Collection

accidentally rammed by the *Lackawanna*. The huge hole in the *Hartford* was only a few inches above the water line. The lone Confederate vessel was surrounded by the sloops and monitors. After the *Tennessee's* steering chains and smoke stack were shot off, the vessel was impossible to steer and the ship filled with smoke. The ram was forced to surrender at about 10:00 A.M. after the hour-long engagement.

Admiral Buchanan suffered a broken leg during the engagement. The Confederate casualties were 12 killed and 19 wounded. The Union casualties were 145 killed and 170 wounded. More than half of the Union deaths were the result of the sinking of the *Tecumseh*.

The Defenses of the Mississippi River and New Orleans Louisiana

Fort Livingston, Grand Terre Island Begun 1841

Built on a barrier island, Fort Livingston was erected to guard the back door to New Orleans through Barataria Pass. It is the most westerly of the Gulf of Mexico forts. It was built on top of a large and ancient Indian shell midden, near the site of Jean Lafitte's pirate headquarters. The shells were used in the construction of the fort for brick mortar and for the tabby walls that were faced with brick. The fort was named in honor of Edward Livingston, Louisiana Senator, Secretary of State under President Jackson, and brother of Robert Livingston, one of the negotiators of the Louisiana Purchase.

The current fort is actually the second design. The first design was identical to General Simon Bernard's design for Forts Pike and Macomb, east of New Orleans. Since construction did not start until the early 1840s, Bernard's plan was shelved in favor of General Joseph Totten's design. The final design was very similar in shape to Fort Barrancas, Florida, a large quadrilateral. The size of the fort's counterscarp wall was 407 feet long on each side. Each land front was 288 feet long. Each sea face was 240 feet long. The main work was two-tiered. The channel face and a portion of the northern front contained musket galleries in the first tier. On the top tier guns were mounted en barbette. To fire down the fort's flanks and faces, a counterscarp gallery was built with 8 counter-fire rooms and four magazines; musket embrasures ran the entire length of the fort's land fronts.

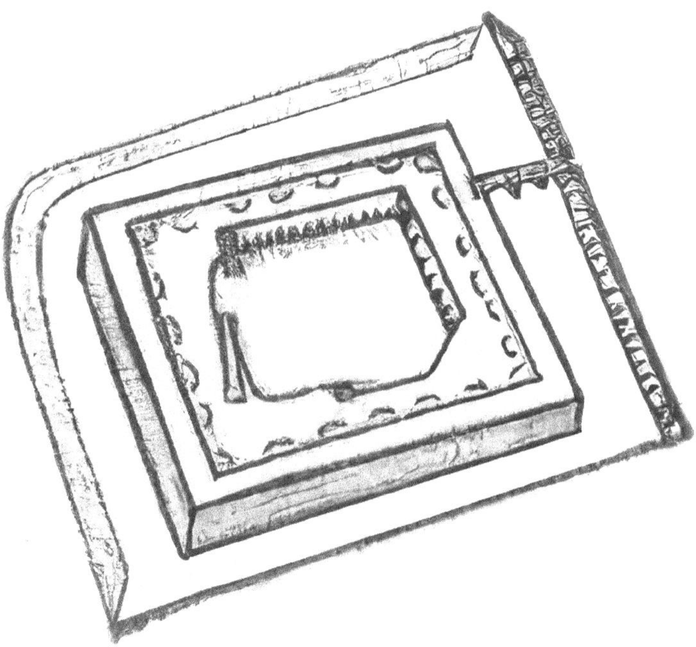

Isometric sketch of Fort Livingston Dale Manuel Collection

Looking toward the end of the fort's channel-side face that once joined with the south wall, one can see the vent of a counterscarp embrasure. A sea-damaged vaulted gallery room is on the right Dale Manuel Collection.

Although the openings look like flank gun embrasures, they are actually vents. Only the very tops of the sand-covered embrasures can be seen in the corner where the two counterscarps meet at an 80-degree angle. John Weaver photo

The channel-side of the parade; on the right corner is the sallyport.
Photo by John Weaver

Musket loopholes can be seen along the fort's channel face. To the left is the flank side that contains the fort's sallyport. Howitzer vents. Mark Berhow Collection

The land side of the parade. On the right are vaulted buttress arches. Photo by John Weaver

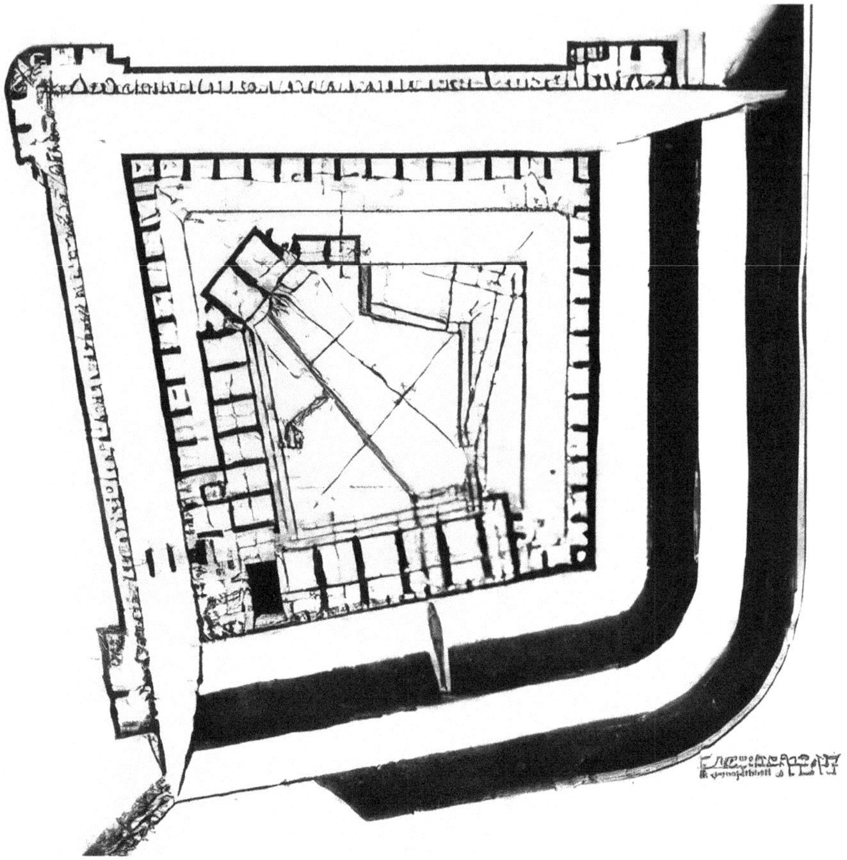

Plan of Fort Livingston. NARA

On three sides of the parade were quarters, a magazine, an artillery storeroom, a sutler store, a bakery, and a blacksmith shop. Two stairs were used to reach the barbette tier. On the southern side of the parade was an earthen ramp and embankment that supported the barbette guns on the terreplein. In order to buttress the weight of the embankment, vaulted arches were built between each support column wall. The southern wall contained three arches between each support column, while the northern wall contained only two arches. The southern wall did not contain galleries or rooms. Three cisterns located beneath the parade supplied water from rain since Grand Terre is a desert island.

The fort's sallyport is located in the north wall. A drawbridge once spanned the dry ditch to provide entrance to the fort. Dry ditches were located on both land fronts. On the channel and sea faces, a wet moat was dug that would fill with sea water. A barracks was once located outside the fort.

The Louisiana State Militia occupied the fort in January, 1861. The Confederates found the fort in a dilapidated condition. Confederate General Mansfield Lovell ordered four companies, totaling 300 men, to garrison the fort. A total of 15 guns were mounted in the fort. The armament included two howitzers, four 12-pounders, seven 24-pounders, a rifled 32-pounder, and an 8-inch Columbiad. The fort was evacuated on April 27, 1862, without action when New Orleans fell to the Federal Fleet. Federal troops reoccupied the fort on February 26, 1863.

After the war, the fort was put on caretaker status. In 1888, it was put in the charge of the lighthouse keeper. In 1856, a 55-foot octagonal brick lighthouse had been built next to the fort. The fort was officially abandoned by the Army in 1892. In 1893, a hurricane breached the southwest corner. Another hurricane destroyed the already weakened south wall in 1915.

The State of Louisiana gained ownership of the site in 1923. Over the years, hurricanes and storms have hit the fort hard and done much damage. The lighthouse is long gone. Currently Fort Livingston is best described as a ruin, but almost three-quarters of the fort still remains. Only the southern wall and earthen embankment have col-

Three musket loopholes in one of the vaulted musket gallery rooms
Photo by John Weaver

An old photo of the ruined end of the eastern counterscarp gallery. The 1897 wood skeleton lighthouse and keeper's house in the background are now gone.
John B. Gasquet Collection, Louisiana State Archives

View from the lighthouse of the fort's sallyport, northern wall, and parade. John B. Basquet Collection, Louisiana State Archives

Entrance to the sallyport Mark Berhow Collection

Channel end of the counterscarp gallery Mark Berhow Collection

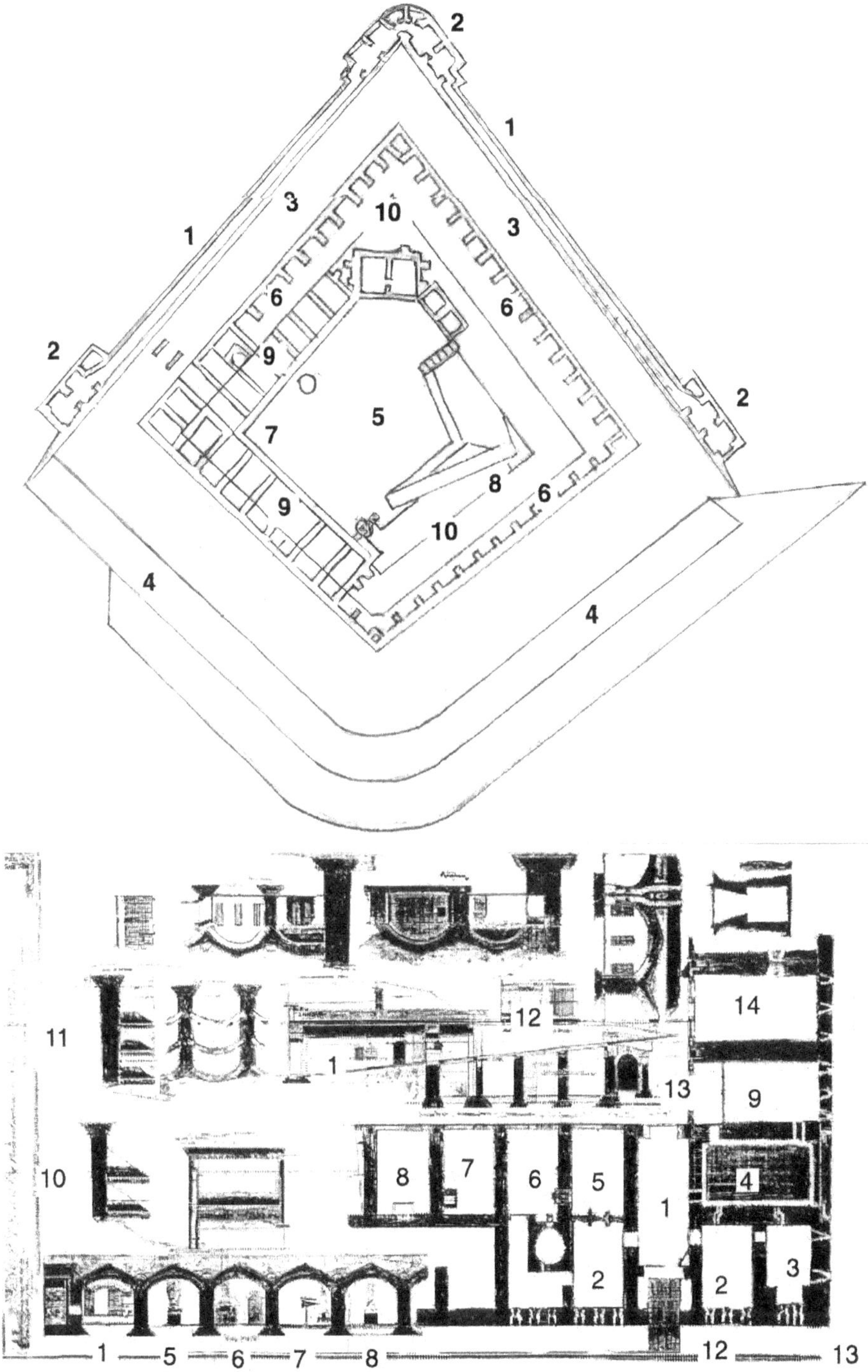

Plan and sections of Fort Livingston. Legend: 1. Sallyport; 2. Guard room; 3. Prison; 4. Magazine; 5. Sutler; 6. Bakery; 7. Blacksmith; 8. Carpenter; 9. Artillery Store; 10. Double Arches; 11. Triple Arches; 12. Drawbridge; 13. Counterscarp Gallery; 14. Musket Gallery Room Dale Manuel Collection

lapsed, leaving the parade open on that side. Most of the west and east walls, as well as most of the counterscarp gallery, are intact. The north wall that contains the sallyport is completely intact. The remains of the fort have been stabilized with a breakwater.

Fort Jackson, Plaquemine Bend, Mississippi River Begun 1822

Located 75 miles southeast of New Orleans, Plaquemine Bend has been the key to the defense of the city since 1746 when the French first built fortifications there. Today, the remains of two brick forts, Fort Jackson and Fort St. Philip, are sinking into the silt of the subtropical jungle of the Mississippi Delta. A sharp turn in the river, and a current of four miles an hour, made this point in the river ideal for fortifications to defend against slow, wind-dependent sailing ships. Named for General and President Andrew Jackson, Fort Jackson was built between 1822 and 1832 to create a crossfire with the older Fort St. Philip across the Mississippi River. The forts are 700 yards apart across the channel. It was here that a mortal blow was dealt to the newly established Confederacy, for it was here that the Federal fleet ran the gauntlet between the forts on its way up river to capture New Orleans which, at that time, was the South's largest and most strategic city.

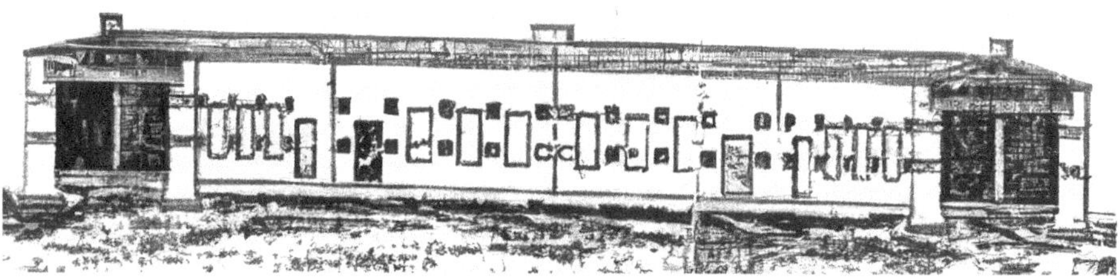

Section of Fort Jackson Citadel (Sketch) Dale Manuel Collection

The fort's parade and channel-side vaulted gun casemates. The parade's center once held a decagonal citadel. Photo by John Weaver

View of one side of the fort, the bastion, and wet ditch Photo by John Weaver

General Simon Bernard designed Fort Jackson in a classical style of French military architecture. The fort is a brick-and-earth pentagon with bastions at each of its five salients. Because of the swampy soil conditions, a grillage of three layers of cypress timbers topped by cypress planks was used as a footing for the foundation of the fort's heavy walls. Each of the five bastions had howitzer embrasures in each flank to control the adjacent curtain. The fort is two-tiered with 16 vaulted casemates in the two curtains bearing on the river. Guns were also mounted *en barbette* on the top tier.

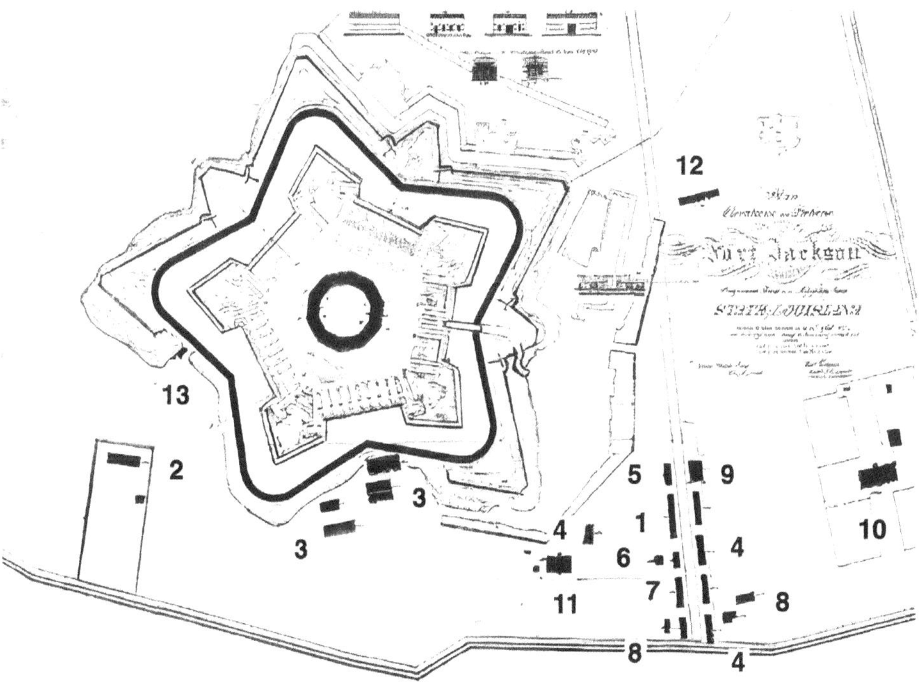

Plan of Fort Jackson and surrounding buildings in 1827. Buildings include: 1. Mess Hall; 2. Convicts' Quarters; 3. Material Sheds; 4. Quarters; 5. Bakery; 6. Engineer Quarters; 7. Blacksmith's Shop; 8. Stores; 9. Hospital; 10. Officers' Quarters; 11. Permanent Quarters; 12. Stable; 13. Steam Engine. Dale Manuel Collection

In the center of the parade was a 10-sided citadel, or defensive barracks, with a perimeter of 500 feet. The citadel could house 400 to 500 men and had two tiers of musket loopholes in both the outer and inner walls. Entrance to the citadel was through a sallyport aligned with the fort's main sallyport. Doors opening to the rooms were in the inner walls.

A landward front of Fort Jackson with the postern in the center of the curtain, flanked by two bastions Photo by John Weaver

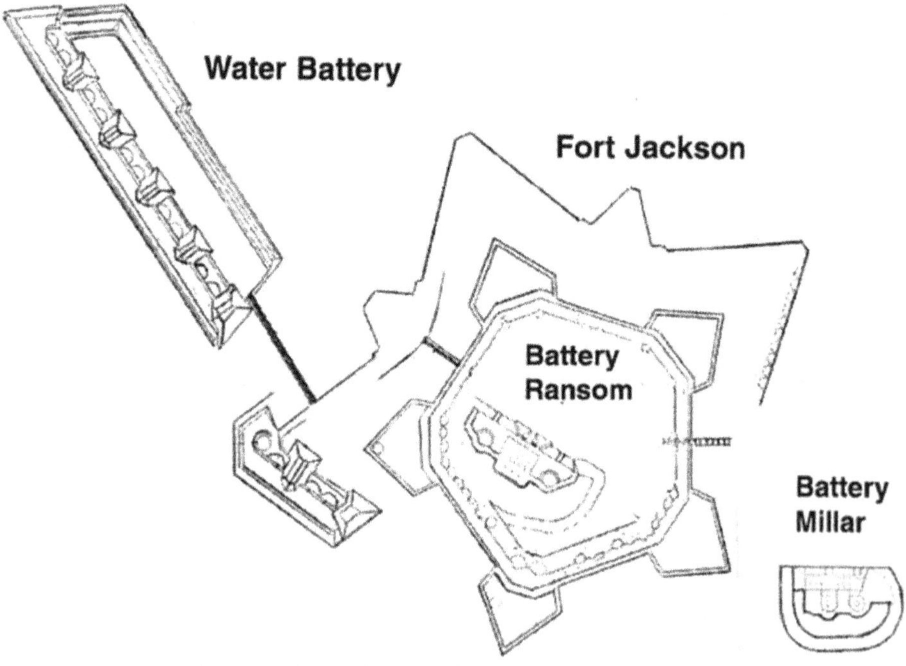

Fort Jackson today with water battery and Endicott batteries John Weaver Collection

The fort had two wet ditches with drawbridges. In the main sallyport, a pulley mechanism was used to raise and lower the bridge. In the fort were furnaces that would heat shot red hot. When fired at attacking wooden ships, the shot could start fires or even ignite the ship's powder magazine. Between the ditches, a brick coverface that contained batteries of guns mounted en barbette was built. Each side of the fort measured 380 feet from the bastion salients. The perimeter of the main enceinte is 2,274 feet. The scarp rose 25 feet above the wet ditch. A water battery of 660 feet by 270 feet was built in time for the Civil War. Major Pierre Gustave Toutant Be-

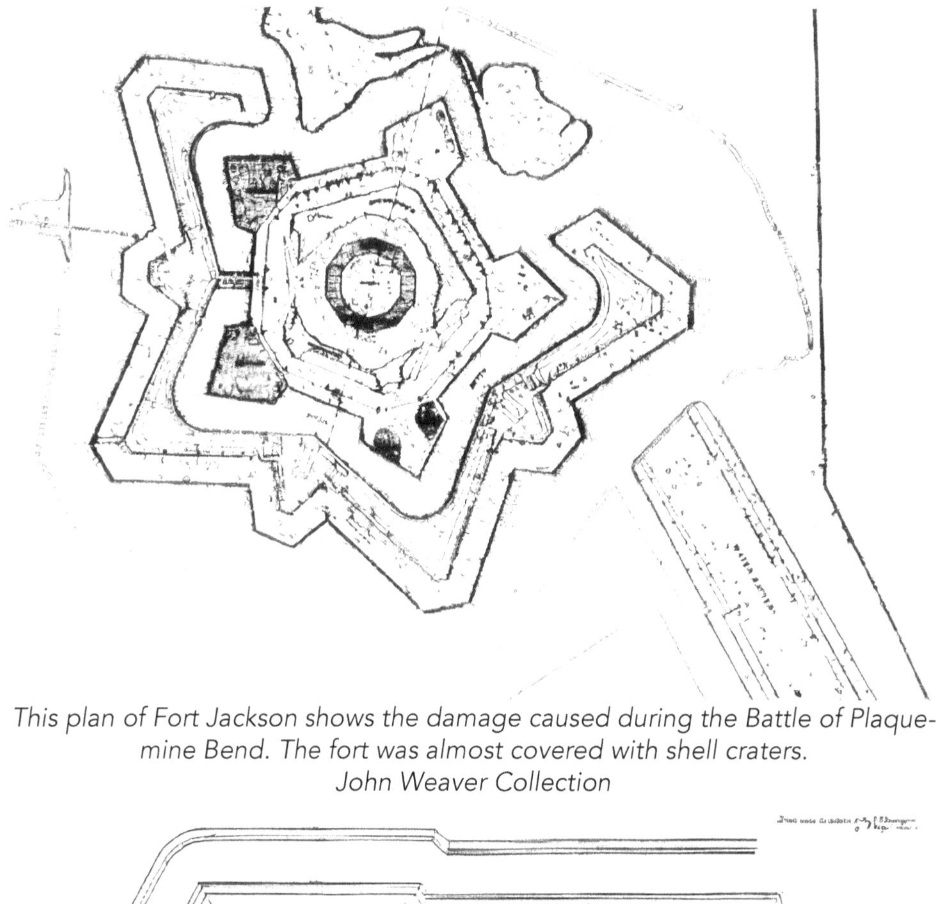

This plan of Fort Jackson shows the damage caused during the Battle of Plaquemine Bend. The fort was almost covered with shell craters.
John Weaver Collection

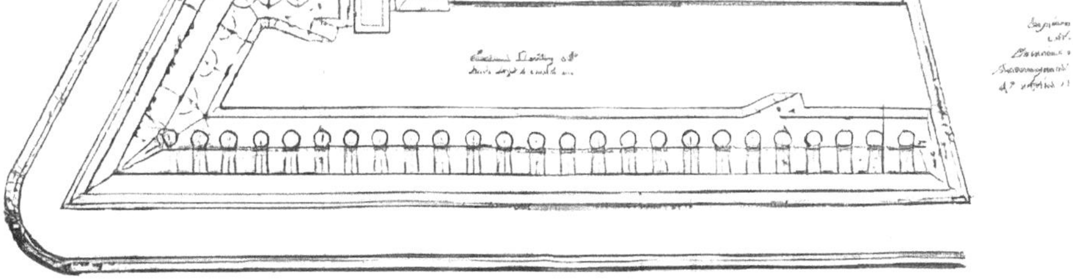

Major P.G.T. Beauregard's plan for the original Water Battery
John Weaver Collection

The interior of a casemate with embrasure. Each gun emplacement had two traverse arcs in the floor. Photo by John Weaver

auregard's plan for this original water battery contained 31 gun emplacements with a magazine. It was replaced in the 1870s with a battery for larger Rodman guns.

The fort was lightly garrisoned until it was seized by Louisiana forces on January 8,1861. During the Civil War, both forts were garrisoned by ten companies totaling nearly 1000 men commanded by Confederate General Johnson K. Duncan. Fort Jackson's armament included one 10- and two 8-inch mortars, ten 24-pounder howitzers, two 48-pounders, one 10- and three 8-inch Columbiads, sixteen 32-pounders, two 32-pounder rifled guns, six 42-and twenty-six 24-pounders. Although the fort was originally planned for 93 guns, only 69 guns were mounted.

A barricade or raft of eight hulks, filled with logs and connected by a chain, was secured to both banks just below the forts. The raft was held in place by the weight of 15 anchors. Above the forts, the Confederates had a small flotilla of 19 vessels of various kinds that included the famed cigar-shaped ironclad Manassas and 11 gunboats. When the Federal fleet entered the Southwest Pass in March, 1862, one of the war's great ship-to-shore battles commenced. At 9:00 a.m., April 18, Commander David D. Porter's fleet of 21 mortar schooners opened fire on the forts. For six days and nights these deadly, 13-inch mortars bombarded the forts with about 18,100 shells. On April 20, crews from the Federal ships Pinola and Itasca succeeded in cutting the barricade chain. On the night of April 24, at 2:00 a.m., Flag Officer David G. Farragut began to run the gauntlet of the two forts with 17 ships armed with 154 guns. As the ships appeared before the forts, there was a thunderous and horrendous crossfire as guns belched out fire and smoke.

Fire barges were sent downriver in an attempt to burn the Union vessels. Above the forts, the Confederate vessels joined in the melee. After four hours of fighting, most of the out-gunned Confederate vessels were destroyed and 13 Federal vessels were steaming their way to capture New Orleans, which surrendered April 29.

Fort Jackson was badly damaged during the battle. Its scarp was cracked, its citadel burned, cisterns damaged, and guns destroyed. The loss of men was relatively light with 9 dead and 33 wounded. Some of the men at Fort Jackson were demoralized from days of bombardment, and fearing a river and land attack against the fort, mutinied on the night of April 27. After the mutiny, the fort was forced to surrender on April 28.

On the parade is Battery Ransom, that once held two 8-inch breech-loading rifles on "disappearing" carriages Photo by John Weaver

View of Battery Millar that once held two 3-inch rapid-fire rifles Photo by John Weaver.

Gun platform of Battery Ransom Mark Berhow Collection

The casemate gallery Photo by John Weaver

After the war, the fort served as a prison. Between the war and the early 1870s, both Plaquemine Bend forts were strengthened and upgraded with new batteries and armament. The Water Battery was remodeled with 10 gun emplacements and five bombproof magazines. Four gun emplacements and a bombproof magazine replaced batteries on the south side wing. A wing battery was added to the north side. The new armament consisted of 8-, 10-, and 15-inch Rodman guns and 100-pounder Parrott rifles. The fort was eventually put on caretaker status.

During the Spanish American War in 1898, a concrete battery, named Battery Ransom, with two 8-inch breechloading steel rifles, was built on the fort's parade. The guns of Battery Ransom were Model Ml888 and were mounted on Model M1898 "disappearing" carriages. The guns were in service from 1899 through 1918. Another concrete battery with two 3-inch rapid-fire rifles, named Battery Millar, was built next to the river on the north side of the fort. Battery Millar's guns were Model Ml898 mounted on 1898 masking parapet carriages. The guns were in service from 1901 through 1920. The fort had a channel-mining depot. There was once a torpedo storehouse, a mining casemate in one of the bastions, and a torpedo loading room near the wharf. The fort saw active duty in WW I and was decommissioned by the Army in 1922.

Fort Jackson and its water battery were restored and was operated as a park and museum by the Plaquemines Parish Commission Council. A nearby levee on the west side of the river protects the site. Pumps keep the fort's wet ditch at the present level. After Hurricane Katrina the interior of the fort has been closed to the public.

Fort St. Philip, Plaquemine Bend, Mississippi River Begun 1793

Built on the site of Fort Plaquemine (1792) by the Spanish, the new brick fort was named Castillo San Philipe in honor of the King of Spain. It was built on the east side of the river near the confluence of the Mardi Gras Bayou and the Mississippi River, and was placed opposite Fort Bourbon to provide crossfire. The United States gained possession of the fort in 1803 with the Louisiana Purchase.

Sketch of Fort St. Philip Dale Manuel Collection

During the War of 1812, Fort St. Philip was remodeled and enlarged. The armament consisted of two 32-pounders, one six-inch gun, one 8- and one 5 ¼ -inch howitzer, one 13-inch mortar and 29 24-pounders. On January 9, 1815, a British fleet attacked the forts. The bombardment continued until January 18. During the bombardment, the fort sustained damage from hundreds of shots. Only one of the British vessels was damaged. Still, the garrison held firm and the British vessels withdrew upon hearing of the British defeat at Chalmette. The fort was lightly garrisoned after the war until 1831, when it was abandoned. Repairs were made and new armament was installed in 1842. For years afterward, the fort was on caretaker status.

Fort St. Philip is a two-tiered, irregular parallelogram with bastions and bastionettes. The size of the main enceinte at its greatest points are approximately 360 feet by 500 feet. In the first tier are vaulted galleries with vaulted musket loopholes. These loopholes were to sweep the curtains with flanking fire. Most of the fort could be swept by the musket loopholes or by guns mounted en barbette on the top tier. Two main bastions on the river fronts directed fire down the river channel.

The fort was seized by Louisiana State Troops in 1861. The Confederates mounted a total of 46 guns in the fort. The armament consisted of one 10- and two 8-inch mortars, ten 24-pounder howitzers, two 48-pounders, one 10- and three 8-inch Co-

A small bastionette (right center) that was at the end of the gorge wall (right). The wall on the left contained a first-tier musket gallery that is now almost completely covered by silt and the remains of the wet ditch. Dale Manuel Collection

The parade and barbette emplacements that ran the perimeter Photo by John Weaver

River side curtain and main bastion. Silt has almost covered the musket loop-holes that were in these walls. Photo by John Weaver

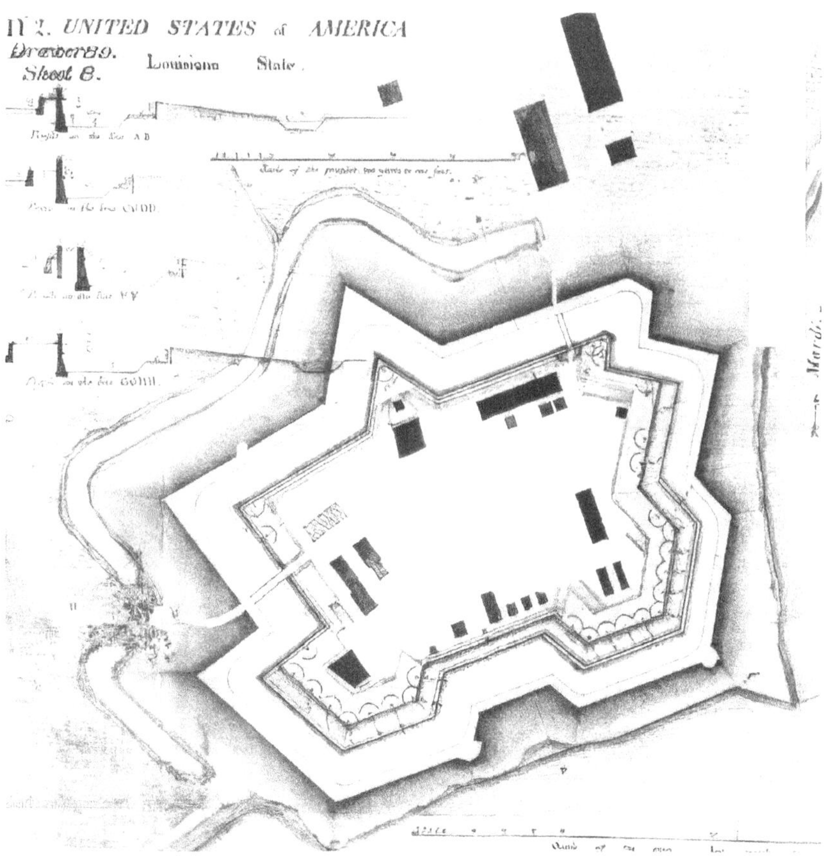

Plan of Fort St. Philip Dale Manuel Collection

lumbiads, sixteen 32-pounders, two 32-pounder rifled guns, six 42-pounders and twenty-six 24-pounders. Batteries were built on each side of the fort. The unfinished Confederate ironclad ship Louisiana was towed to Fort St. Philip to serve as a floating battery.

Little damage was done to Fort St. Philip during the battle with the Federal fleet on the night of April 24. Of the fort's garrison, there were two dead and four wounded. The most Confederate casualties from the Battle of Plaquemine Bend were 73 dead and 73 wounded of the outgunned flotilla.

The Rebel flotilla could not stop the progress of the Federal fleet, despite the valiant efforts of Lieutenant Warley in the Manassas, the mortally wounded Lieutenant

The sallyport and the amount of silt that almost covers it. Photo by John Weaver

The main sallyport is half-buried and still contains iron pulleys that once raised and lowered the drawbridge. Photo by John Weaver

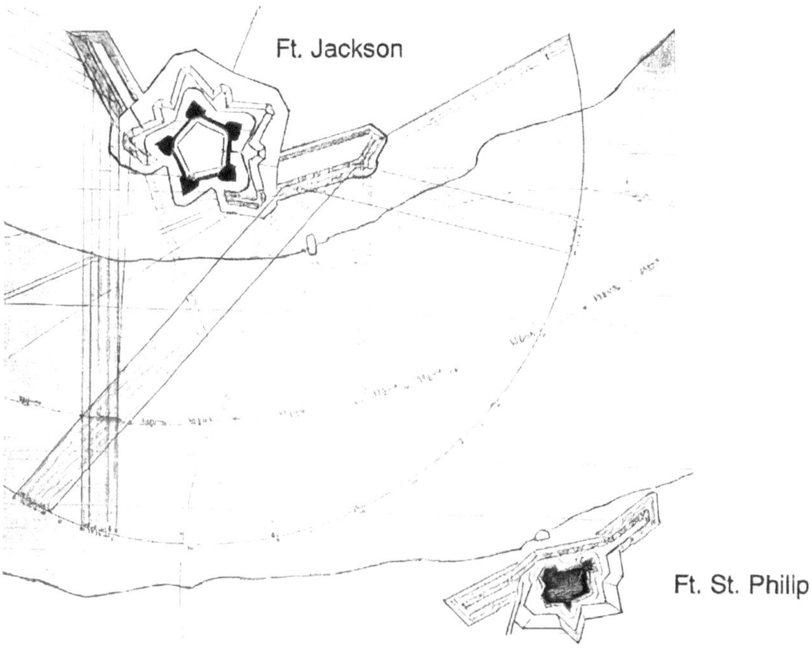

After the Civil War wing batteries were added to both the Plaquemine Bend forts. The armament was upgraded with Rodman guns and Parrott rifles. Dale Manuel Collection

Huger in the McRae, and Lieutenant Kennon in the Governor Moore. The Federal loss was 37 killed and 147 wounded. The Union Navy lost four vessels during the fighting. The lead gunboat, Cayuga, took 42 hits but passed the forts.

Unfortunately for the Confederates, two huge ironclads, the Louisiana and the Mississippi, were unfinished at the time of the battle. Both ships could have turned the tide against the Federal wooden ships. Both ships were blown up after the battle to prevent their capture by the Union fleet.

During the upgrade, earth-covered bombproof magazines were built in the batteries of both forts Photo by John Weaver

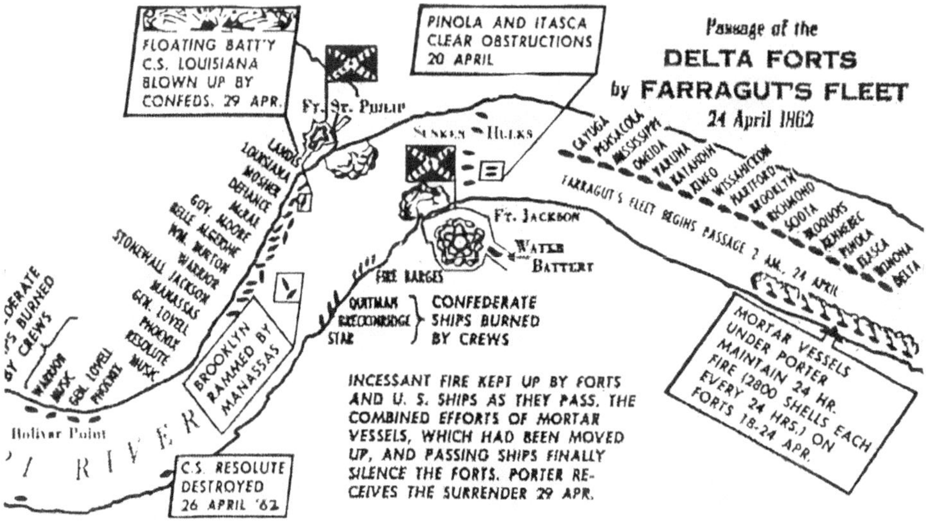

Map of the Battle of Plaquemine Bend that spelled the loss of New Orleans and possibly the war for the Confederacy Dale Manuel Collection

After the war, the fort received new armament and batteries. Batteries and bomb-proof magazines were constructed on the river front with wings on both sides. Plans called for 29 gun emplacements to be constructed in the batteries. At least 25 were built to mount 15-inch Rodmans. The fort was eventually put on a caretaker status until the late 1880s.

A gun emplacement at Fort St. Philip. A pintle block is near the center of the traverse arc. Wall indentions gave clearance to the front end of the carriage. Dale Manuel Collection

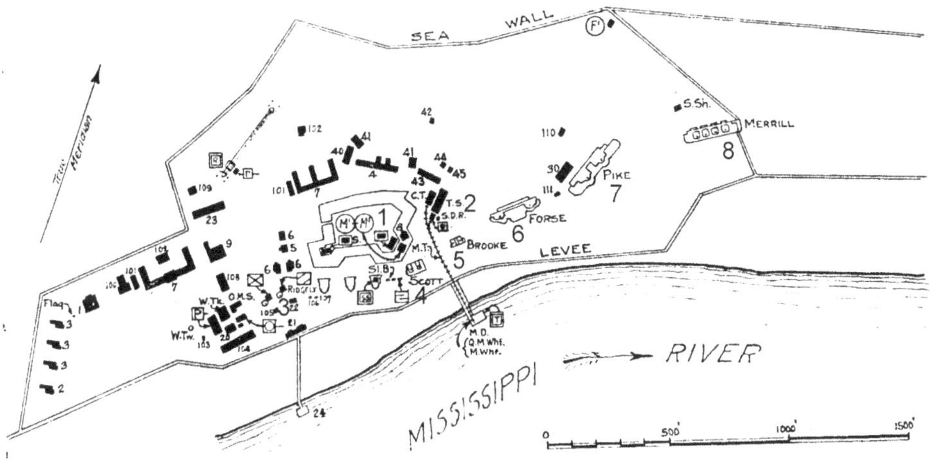

1. Old Fort; 2. Mine Storage; 3. Battery Ridgely; 4. Battery Scott; 5. Battery Brooke; 6. Battery Forse; 7. Battery Pike; 8. Battery Merrill

Fort St. Philip developed into a larger post than Fort Jackson during the Endicott period of coastal defense. It had six concrete batteries, barracks, quarters, stores, hospital, offices, shops, mess halls, bakery, stable, two wharfs, post exchange, a seawall on the back and sides, and a levee on the river side.

The almost-buried entrance to the gorge postern Photo by John Weaver

The parade side of the main sallyport Photo by John Weaver

In 1898, during the Spanish American War, Endicott-era concrete batteries were constructed at the fort. Battery Ridgely mounted two 4.7-inch rifles that were in service from 1899 to 1913. Battery Scott held two 3-inch rifles that were in service from 1901 to 1920. Battery Brooke mounted two 3-inch rifles that were in service from 1904 to 1920. Battery Forse held two 8-inch disappearing rifles that were in service from 1899 to 1918. Battery Pike mounted two 10-inch disappearing rifles that were in service from 1898 to 1919. Battery Merrill held four 6-inch rapid-fire rifles. The guns of Battery Merrill began service in 1907. Two were removed before WWI and two guns remained until 1920.

Located on the south side of the old fort was the center for channel mining operations. A cable tank, mine storage, and dynamite room were connected to the mining wharf by a tramway. The primary mine command station was located on top of the parapet of the main bastion of the old fort and the mining casemate was nearby. Fort St. Philip was an active artillery post until 1904. It was reactivated in World War I. The fort was decommissioned in 1922.

Fort St. Philip is on private property and can only be reached by boat. The remains of the fort are in surprisingly good shape. The only major blemishes visible are a 15-foot hole in the riverfront curtain wall and the removal of the small bastionette that was once attached to the middle of the gorge. Because the east side of the river has no levee, years of flooding have silted in most of the wet moat and almost covered the first tier. Most of the fort's parade has been filled in and the only structure that remains is the ruin of the old forge.

Proctor's Tower, Proctor's Landing, Lake Borgne Begun 1856

Proctor's Tower was built on the shore of Lake Borgne to protect the town of Proctor's Landing and prevent its railroad grade from being used as a route for the invasion of New Orleans. Because soon-to-become Confederate General Pierre Gustave Toutant Beauregard was the supervising engineer, the fort is sometimes referred to as Fort Beauregard by locals.

The gun tower was designed to be a brick rectangle with three tiers. It was designed along the lines of the tours-modeles gun tower favored by Napoleon I for the coastal defense of France. Similar towers were erected at the East and West Martello Towers of Key West, Florida; and Forts Winthrop, Massachusetts, and Porter, New York. In the first tier, large musket loopholes encircled the perimeter of the main work. In the gorge, a small portico gave entrance to the sallyport. On the triangular pediment of the portico is the date 1856. A drawbridge mechanism was to have raised and lowered the gate.

Inside the first tier was a magazine and stairs leading to the top floors. The second tier of the fort was designed to contain 8 casemated gun emplacements and musket loopholes. In order to support the tremendous weight of the guns, the tower was one of the first structures in this nation to use rolled iron beams. These iron girders were used between floors for structural support. Because of the intervening Civil War, the second tier was never finished. Had the top tier been finished, it would have mounted guns en barbette. A rainwater collection system on the roof would have piped water to four cisterns below the first-tier floor. The designed height of the scarp wall would have been about 36 feet had it been completed.

Picture of Proctor's Tower
Gorge area of the tower. Visible is the sallyport's portico, part of the surrounding batteries' tabby revetment wall, and remains of rooms on right
Photo by John Weaver

View of Proctor's Tower
The tower was one of the first forts to use rolled iron girders for support of the second-tier floor. Visible are musket loopholes and fireplaces in the wall.
Photo by John Weaver

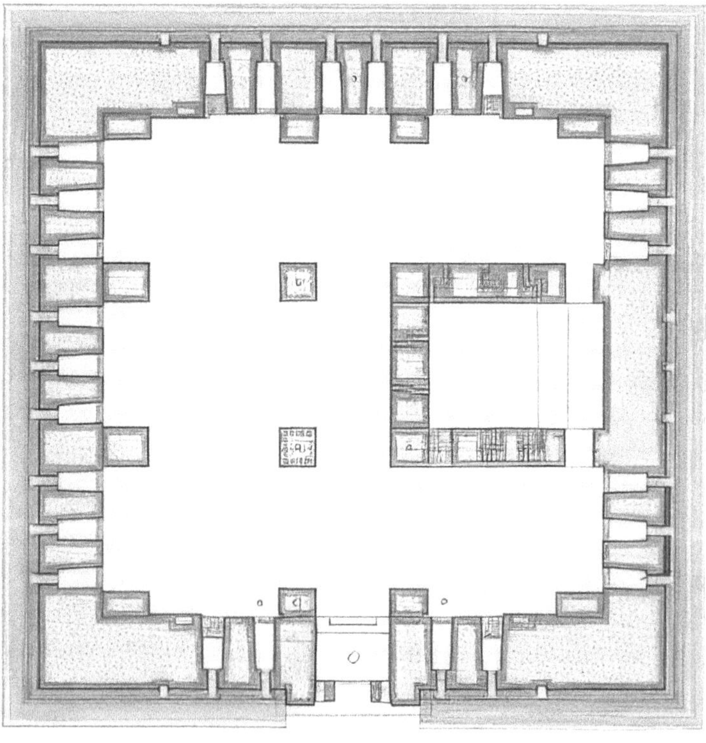

Plan of Proctor's Tower. NARA

Main hall inside the tower. Riveted iron girders are overhead, and the first-floor magazine is on the right. Photo by John Weaver

An outer rectangular battery surrounded the tower with a tabby revetment wall that was two feet thick and mounted guns en barbette. Semicircular gun emplacements were at two corners of the battery. Guard rooms and quarters were built into the battery wall in the gorge area. Behind the fort reservation, wooden barrack buildings were built; these have long since disappeared.

During the Civil War, Confederate General Lovell stationed a company of men with six guns at Proctorville. The wooden bar racks and buildings of the unfinished tower were used by the garrison. The area was evacuated by Confederate troops after the surrender of New Orleans on April 29, 1862.

After the war, the tower's second tier remained unfinished and the post was abandoned by the army. The fort that was once on dry land is at the edge of Lake Borgne. A protective rock breakwater has been added on three sides of the structure. The iron girders that were to have supported the second tier floor have rusted over the years. Iron pipes that were to have carried rain water to cisterns below can be seen reaching skyward. Proctorville and even Old Shell Beach have disappeared. The site has been cut off from the mainland by the digging of the Gulf Outlet Canal and can only be reached by boat.

Tower Dupre, Bayou Dupre, Lake Borgne Begun 1828

This brick, hexagonal, two-tiered tower no longer remains; it was destroyed by Hurricane Katrina. It was built to protect the entrance to Bayou Dupre at Lake Borgne. Though the original design was for three tiers, because of storm damage to the roof and top tier, the fort was modified, and the top story was removed about 1843. The original 34-foot high walls were cut down to 29 feet. Each of the six sides were 24 feet wide. First developed in Corsica, the idea of the Martello tower was to give a 360-degree field of fire for the guns. This arrangement would protect the entire circumference of the fort. Tower Dupre was one of less than a handful of these normally circular towers that were also erected at New Castle, New Hampshire; James Island, South Carolina; and Tybee Island, Georgia.

On the first tier, four of the six sides contained five musket loopholes. In the back side there was one musket loophole on each side of the fort's sallyport. On the second tier, there were eight gun embrasures. Each side had one center gun embrasure except the front side which had one at its center and one at each corner. The second-tier deck was constructed of wooden planks and cross members supported by the walls of the tower and a central support pillar. The top of the structure was roofed and on top was a cupola and a flagpole. A six-gun earthen battery was forward of the tower for added firepower. The tower and battery were surrounded by a wet ditch, filled by the bayou.

The lower tier of the tower was subdivided into rooms. The sallyport gave entrance to a foyer with musket loopholes on its outer sides and stairs. Traveling in a counter-clockwise direction through the rooms of the first tier, a door is entered into

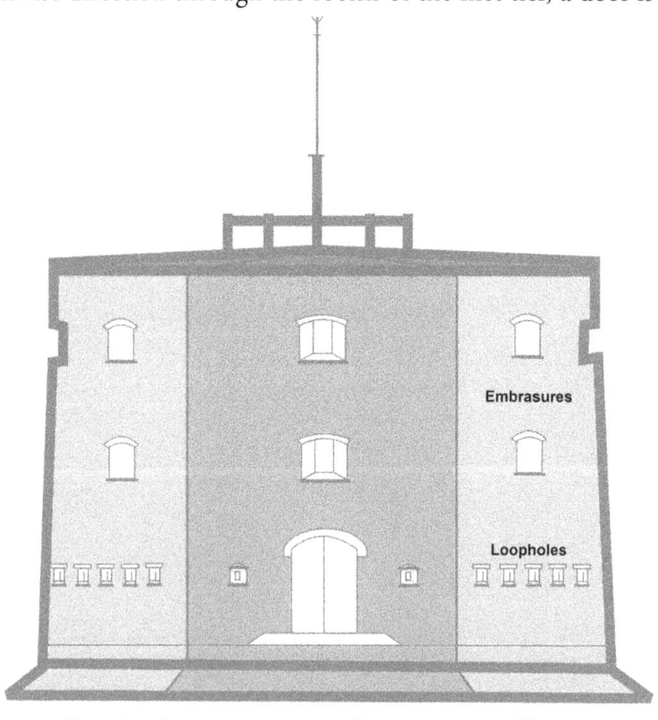

Sketch of Tower Dupre John Weaver Collection

The hexagonal tower Photo by John Weaver

View of the tower's first-tier musket loopholes, sallyport, and second-tier gun embrasures. The tower has been roofed to serve as a fishing camp. Photo by John Weaver

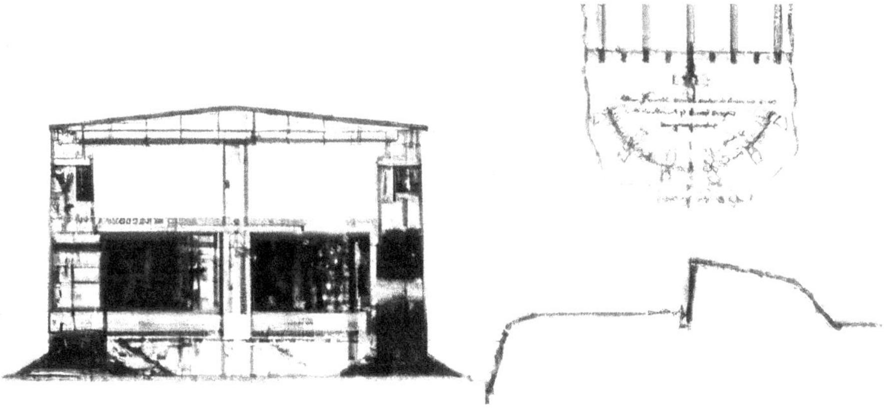

Plan and sections of the original three-tiered tower Dale Manuel Collection

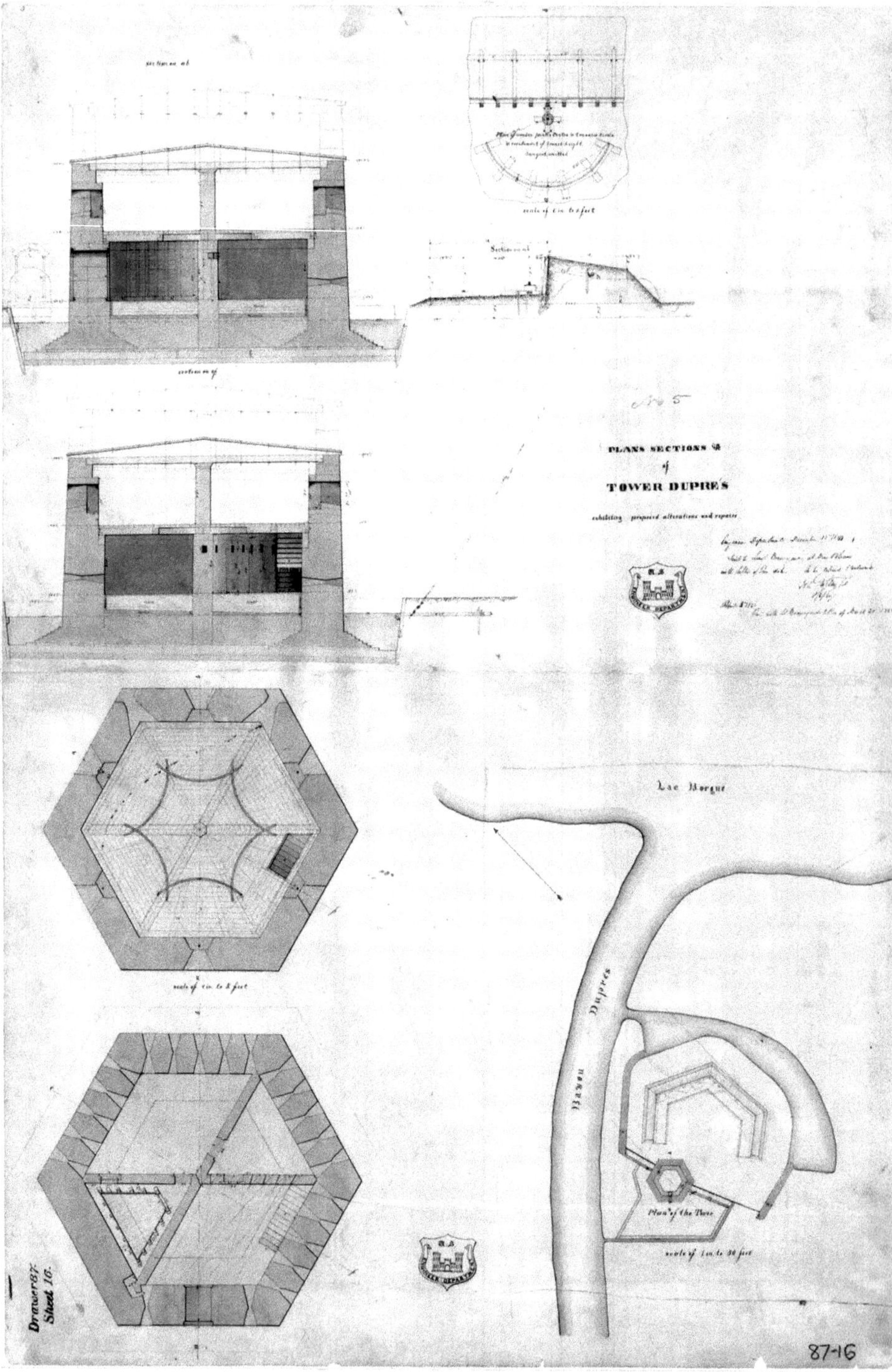

Plans of the modified two-story Tower Dupre and battery
Dale Manuel Collection

Tower Dupre was once on dry land but the ruins are now surrounded by water. It is located at the confluence of the Violet Canal and Lake Borgne. Photo by John Weaver

an adjacent room. The dividing wall of this room and the central pillar contained musket loopholes to fire on the sallyport and prevent entrance by an enemy. The other rooms contained musket loopholes on their exterior sides as well. In the last room that was adjacent to the foyer, but separated from it by a solid wall, was the vented powder magazine.

During the Civil War, the Confederates stationed a company of men at the tower and mounted five 24-pounder guns. The Confederates evacuated the fort when New Orleans surrendered. The ruins of the tower, once on dry land, is in the middle of the Violet Canal where it empties into Lake Borgne. The site is privately owned.

Battery Bienvenue, Bayou Bienvenue Begun 1828

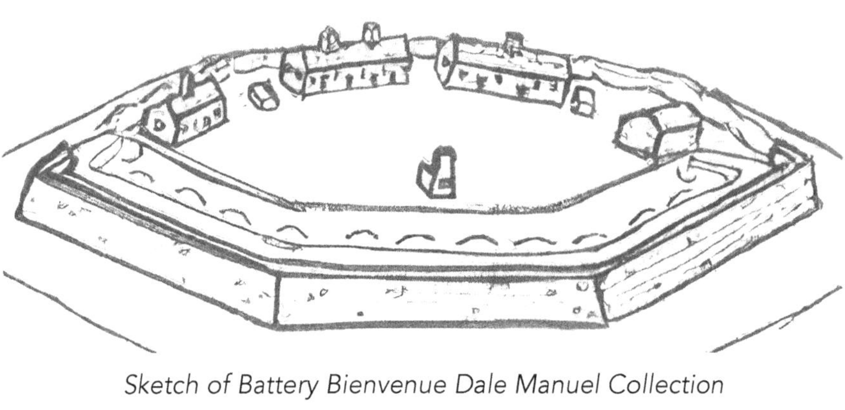

Sketch of Battery Bienvenue Dale Manuel Collection

Battery Bienvenue is located at the confluence of Bayou Bienvenue and Bayou Maxent (Villere), the same route used by the invading British in the War of 1812 to reach the banks of the Mississippi below Chalmette. Battery Bienvenue was a large, rectangular redoubt that was built on the site of the War of 1812 Fort Villere.

The fort is an open battery with a brick scarp in front. The scarp wall is 140 yards long from end to end, with two 155 degree angles that gave it three fronts and angles of fire. The center face is 80 yards long with a face of 30 yards on each side. Behind this brick scarp were mounted guns en barbette on an earth rampart with a brick revetment. The battery was designed for an armament of 20 guns and mortars.

Early plans show that the sides and gorge were enclosed by a wooden palisade, an unusual feature for a Third System work. Likely, there was not that much concern for a rear attack with a site that was surrounded by marsh. Two guns mounted on wooden platforms guarded the gorge palisades. The site was surrounded by a wet ditch. A causeway ran from a landing on the bayou through the wet ditch to a drawbridge and gate in the southern side of the palisade.

On the parade were many structures. At the center of the front battery was a brick hot-shot furnace. On the north side of the parade is a strong, square magazine measuring 24 feet on each side. On the gorge side of the parade were two brick quarters, one with two rooms and one with four rooms. Each of the quarters measured 26 feet by 90 feet; at each end of these quarters were cisterns. On the south side of the parade was a storage building with two rooms measuring 24 feet by 45 feet.

The front brick scarp angle. This face fired directly down the bayou
Photo by John Weaver

The strong square magazine is viewed from the ruins of the gorge quarters
Photo by John Weaver

A cannon mounted on the brick and earth ramparts is aimed toward the bayou
Photo by John Weaver

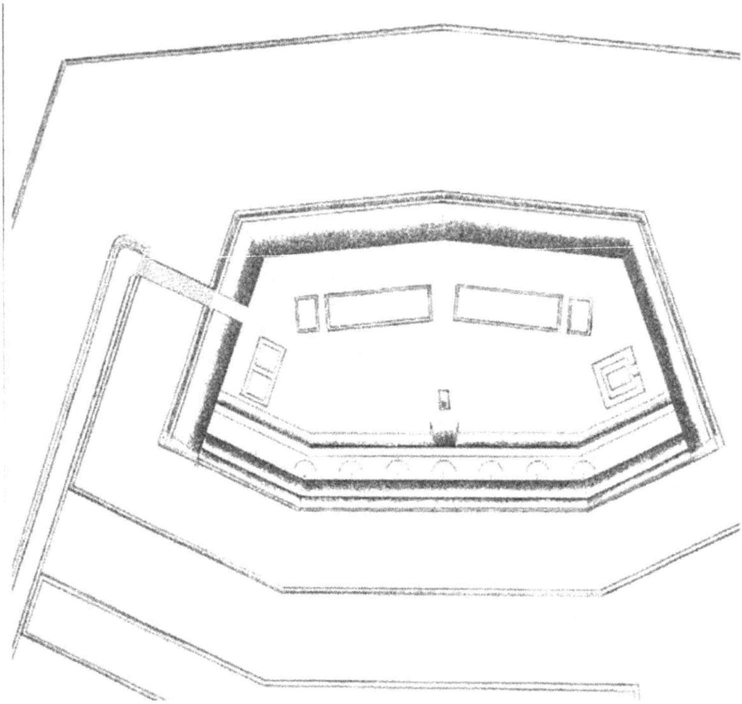

Original plan of Battery Bienvenue. NARA

Remains of the hot-shot furnace at Battery Bienvenue. The tree line marks the ruins of the gorge barracks. Mark Berhow Collection

During the Civil War, the Confederates garrisoned the battery with a company of men armed with 10 guns. The Confederates evacuated the fort with news of the fall of New Orleans to Federal troops. After the war, the army abandoned the installation.

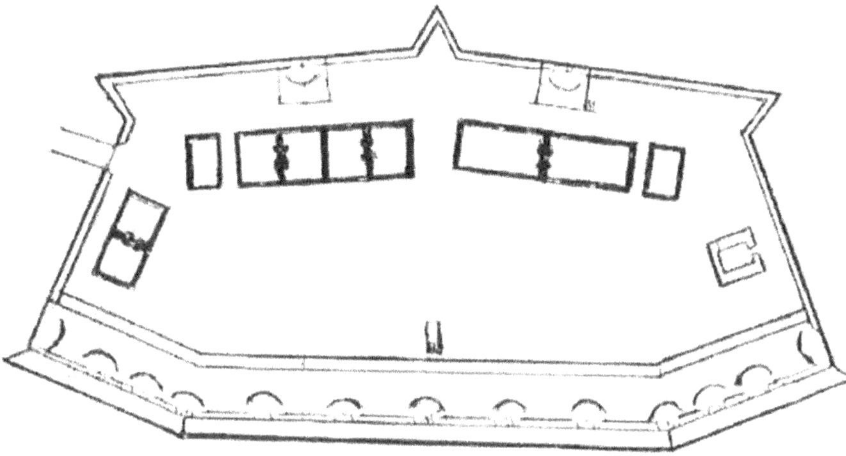

Plan of the battery with stockade palisades around the sides and gorge. Two wooden gun emplacements were added to the gorge palisade Dale Manuel Collection

The end of the scarp Mark Berhow Collection

The magazine door opening Mark Berhow Collection

The site is on private land and can only be reached by boat. Most of the brick scarp of the front battery remains. Still on the ramparts are six 24-pounder guns of 1819 and 1839 models. One of these guns is dated 1832, one is dated 1837, one is dated 1839, and the other three are dated 1859. The brick magazine is still intact. The hot-shot furnace is in ruins, as are the quarters on the gorge side. The cisterns are still visible. The storage building has collapsed except for an outline of the walls that stands three feet above the present level of the site. Because of subsidence, the battery's parade is the same level as the surrounding marsh. The former wet ditch remains on the three faces of the battery.

Jackson Barracks, Orleans Parish Begun 1834

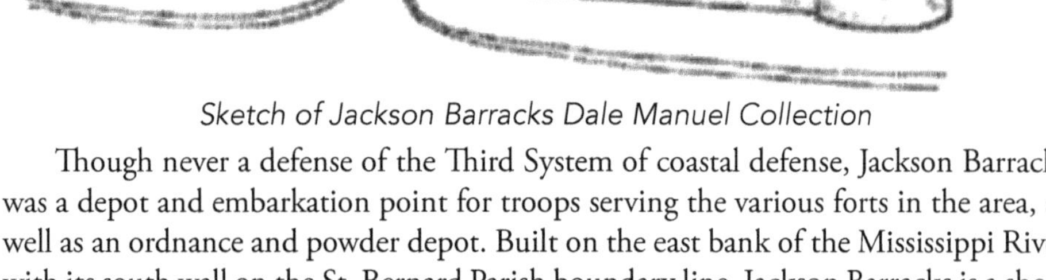

Sketch of Jackson Barracks Dale Manuel Collection

Though never a defense of the Third System of coastal defense, Jackson Barracks was a depot and embarkation point for troops serving the various forts in the area, as well as an ordnance and powder depot. Built on the east bank of the Mississippi River with its south wall on the St. Bernard Parish boundary line, Jackson Barracks is a short distance from the site of the Chalmette Battlefield. The defensible barracks began in 1834 as a large brick quadrangle measuring 560 by 1500 feet, with round towers

A round tower was located at each corner of the quadrangle. Two flanking gun embrasures were on the bottom floor of each tower. Mark Berhow Collection

One of the barracks in the cluster of four within the quadrangle Mark Berhow Collection

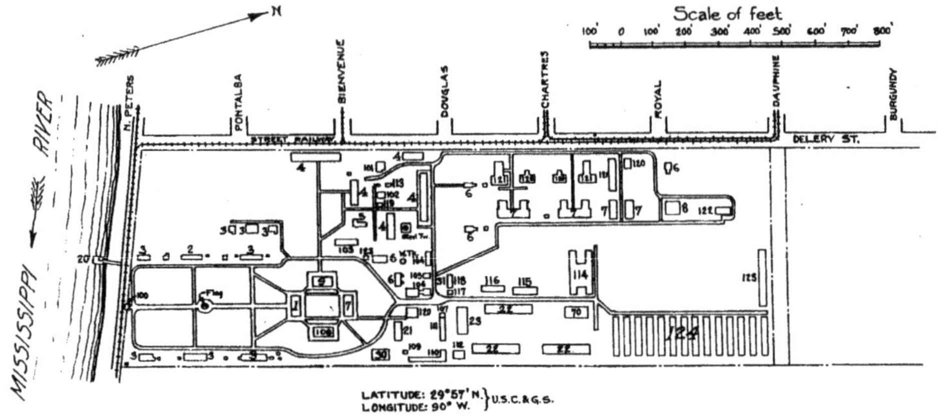

Plan of Jackson Barracks NARA

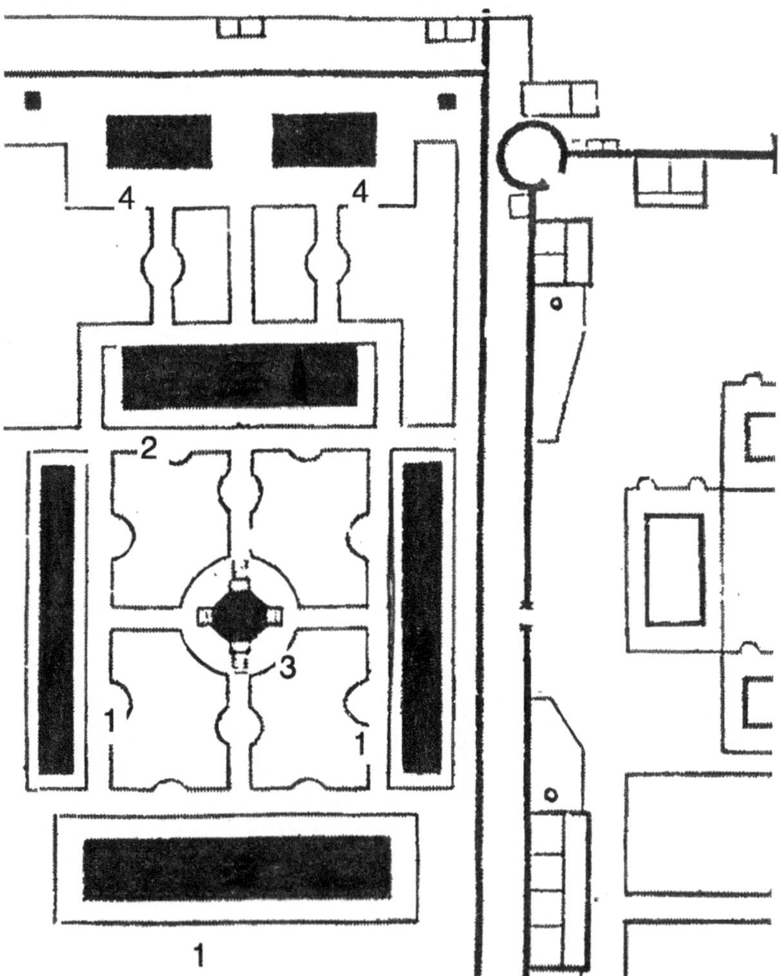

Plan of the hospital: 1. Hospital Wards; 2. Nurse's Quarters and Mess Hall; 3. Dispensary; 4. Kitchens and Quarters Dale Manuel Collection

Parade Mark Berhow Collection

on each of its four corners. The three-tiered towers were crenelated on top and had loopholes for musket fire on a lower tier. The top of each tower mounted a small gun to sweep two sides, much like a Martello tower, and two embrasures on the bottom tier added flanking fire as well. On the spacious parade is one cluster of four barrack buildings as well as singular barracks along the walls of the quadrangle. The buildings are some great examples of Greek revival architecture. The walled U.S. magazine was positioned 2700 feet to the rear of the quadrangle. This was done for safety reasons as lightning can ignite gunpowder.

The post served as an embarkation point for troops leaving for Florida's Second Seminole War and for the Mexican War. In 1845, Lt. U.S. Grant served here awaiting transport to Texas. The regulars were housed in the Barracks, while the volunteers

Walled U.S. magazine is now the post museum Mark Berhow Collection

This sallyport was in the north wall of the quadrangle and led to the hospital area. When open, the twin gates occupied the indentations on each side. Mark Berhow Collection

camped in tents at the damp Chalmette Battlefield. The volunteers must have envied the regulars in their comfortable quarters. During the Mexican War, the post was used as a hospital.

In 1849, a hospital complex was erected outside the north wall. Four large wooden two-story buildings formed a square. Three of these buildings were wards and the

One of two flank gun embrasures in each of the flank towers. Flank guns could sweep the sides of both walls. Dale Manuel Collection

fourth was a mess hall and nurse's quarters. In the center of the square was a two story octagonal building that served as a dispensary and office.

Seized by Louisiana troops in January, 1861, it was abandoned by the Confederates and reoccupied by Federal troops on May 1, 1862, after the Federal fleet passed the Plaquemine Bend and Chalmette forts.

In 1907, Fort Jackson and Fort St. Philip became sub-posts of Jackson Barracks. Various coastal artillery units were stationed at the lower forts for target practice.

The Barracks once had a two story commanding officer's quarters with a built-in sallyport that led to a wharf on the river, as well as two additional round towers. These structures were removed in 1912 to make way for a new levee.

During World War II, U.S. troops were stationed at the barracks. The state took ownership in 1946. The 85-acre reservation is operated by the Louisiana National Guard. Two towers, the powder magazine, and fourteen original buildings remain on this well-kept and historical post.

Fort Macomb, Chef Menteur Pass, Lake Pontchartrain Begun 1820

The near-twin of Fort Pike was originally named Fort Wood. It was named for Colonel Eleazer Wood, who was killed near Fort Erie, Ontario, in the War of 1812. In 1851, the designation of Fort Wood was moved to the star shaped fort on Bedloes Island that now forms the base of the Statue of Liberty in New York Harbor. The brick fort, built to protect the Gentilly Road and the Chef Menteur Pass between Lake Borgne and Lake Pontchartrain, was renamed Fort Macomb, for a veteran of the War of 1812, and member of the Army Corps of Engineers, General Alexander Macomb.

The fort has a sector shape, two sides of a triangle and a curved third side. The semicircular water face contains 13 vaulted casemated gun rooms on the first tier and eleven guns en barbette on the second tier. Each of the two sides has a demibastion on the water face end. Both sides come together to form a full bastion on the landward side. In order to achieve flanking fire, each side has a gun embrasure in the curtain angles of the demibastions and in the curtain angles of the full bastion. The curtain of

Fort Macomb from the other side of Chef Menteur Pass Photo by John Weaver

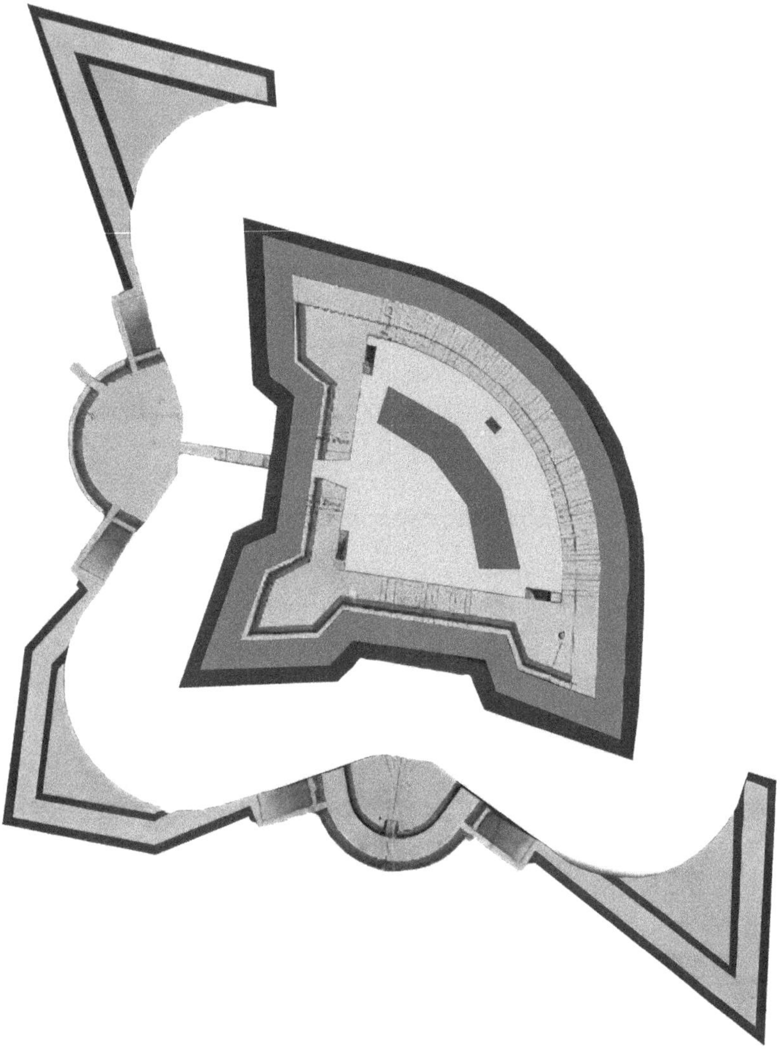

Plan of Fort Macomb John Weaver Collection

the south side has three gun embrasures. The curtain on the north side contains the fort's sallyport and two gun embrasures. On the fort's parade is an angled, two-story citadel or defensive barracks 47 by 140 feet. The four-foot-thick citadel walls contain musket loopholes in two tiers to fire at all angles. The parade contained hot-shot furnaces and cisterns. Ramps and stairs lead from the parade to the terreplein above.

The fort had two concentric wet ditches with a coverface and earthen glacis between the ditches. Both the outer ditch and the sallyport had drawbridges. The re-entering place of arms had a demilune with a curved brick passageway that would prevent direct fire on the sallyport.

During the Civil War, the Confederates garrisoned the fort with 250 men and armed it with 30 guns; 21 of them were 24-pounders. The Confederates evacuated the fort after the surrender of New Orleans to Federal troops. After the war, the fort was on caretaker status until 1871.

Some of the damage to the pass face caused by the wakes of boats Photo by John Weaver

Ruins of the curved passage through the covered way that led to the sallyport Photo by John Weaver

View of the wet ditch, sallyport, and ruins of the brick passageway. The ditch has since been filled with earth. Photo by John Weaver

The fort is owned by the State of Louisiana but is not open to the public. A rubble levee now surrounds the fort to prevent damage to the scarp by wave action and the wake of boats traversing the pass.

Fort Pike, Pass Rigolets, Lake Pontchartrain Begun 1819

The first of the Permanent (later called "Third") System forts to be constructed in the United States, Fort Pike, the near-twin of Fort Macomb, was built to protect the Rigolets Pass between Lake Borgne and Lake Pontchartrain. It is named in honor of General Zebulon Pike, explorer of the Western Louisiana Purchase, of Pike's Peak fame, and conqueror of Fort York, Ontario. He was killed at Fort York, now Toronto, in the War of 1812. Designed by General Simon Bernard (1779-1839), it was of the classical style of French engineer Sebastian Vauban. The defense of New Orleans was of primary importance to the Army Corps of Engineers. The Corp's highest classification, the first classification, was granted to many of the forts of Louisiana because of this urgent need for defense.

As with most of the forts built around New Orleans, Fort Pike's foundation was laid on a grillage constructed of layers of cypress timbers and planks to hold the weight of the walls on soft, marshy soil. As long as the cypress was submerged in water, it would not rot.

Officers' quarters, a bakery, and storehouses were built near the fort. Later, a carpenter shop, more storehouses, a blacksmith shop, the commissary, a mess hall, and ordnance storehouses were added. A short distance away a hospital was built at the Lake Pontchartrain shore, at the site of Fort Petites Coquilles, the Spanish fort built in 1793 that Fort Pike replaced.

Inside the casemates of the semicircular channel face. Granite arcs on the floor were used to roll heavy guns in order to aim them. Vaulting supported the weight of the guns on the above terreplein batteries. Photo by John Weaver

The fort's designed armament included twenty four 32-pounder cannons, 15 24-pounder cannons, two 10- and four 13-inch mortars. The planned garrison was for 80 men in peacetime and 400 in time of war.

Semicircular pass face of Fort Pike Photo by John Weaver

The citadels of Forts Pike and Macomb are the only remaining in the U.S. Photo by John Weaver

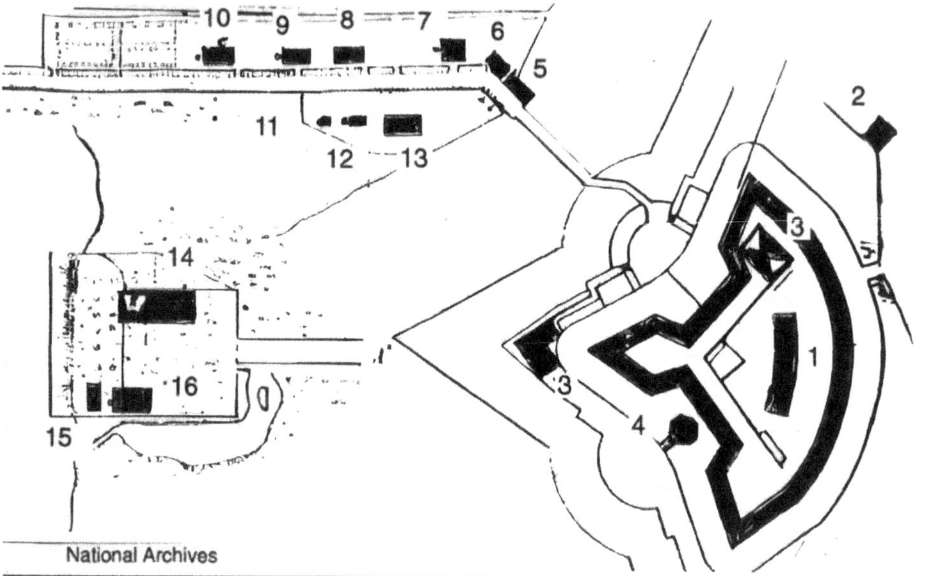

Plan of Fort Pike and hospital that was built on the nearby site of the old Spanish Fort Peites Coquilles. Legend: 1. Citadel; 2. Wharf; 3. Officers' Quarters; 4. Bath House; 5. Boathouse; 6. Commissary Office; 7. Commissary Stores; 8. Ordnance Stores; 9. QM Stores; 10. Carpenters; 11. Blacksmith Shop; 12. Bakery; 13. Mess Hall; 14. Doctor's Quarters with Kitchen; 15. Kitchen; 16. Hospital NARA

Today's view of the terreplein of the water battery, vaulted casemates, and the citadel Photo by John Weaver

The citadels of Fort Pike and its near-twin, Fort Macomb, are the only surviving citadels in the United States. Originally built as a single story, a second story was added in the 1850s. The retreating Confederates burned the citadel's woodwork, as well as the rest of the wooden buildings, in 1862. The rebuilt citadel burned again accidentally in 1887 and is in ruins today.

During the Seminole Wars in Florida, the fort served as a detention center for Indians being transported to the Indian Territory of Oklahoma. Casemates served as cells. As many as 253 prisoners were guarded by 66 soldiers. During the War with Mexico, the fort was an embarkation point for soldiers.

In this old photo, the ruins of the second story citadel and parade are shown. John B. Gasquet

Wet ditch, curtain, demibastion, and flank gun embrasure Photo by John Weaver

Louisiana State Militia seized the fort from a single caretaker in January, 1861. The Confederates garrisoned the fort with 350 men. The armament consisted of a total of 33 guns that included two 32-pounder rifled guns and twenty 24-pounder smoothbore guns. The Confederates evacuated the fort when New Orleans surrendered to Federal troops on April 29, 1862.

The fort is a State Commemorative Area located east of New Orleans on U.S. 90. It is no longer open to the public – the fate of all forts located in Louisiana!

Fort Massachusetts, Ship Island, Mississippi Begun 1859

The Fort on Ship Island, later unofficially called Fort Massachusetts, was built on the western tip of Ship Island, Mississippi, to guard the Mississippi Sound and prevent the island's deep-water harbor from being used as a staging area for the invasion of New Orleans. As it turns out, the island was used as a staging area for two invasions of New Orleans, one by the British in the War of 1812, and one by the Federals in the Civil War.

Fort Massachusetts was one of the last of the Third System forts to be completed in the United States. It was the most advanced semicircular castle ever built. The design allowed a wide, consecutive field of fire on any ship entering the channel of Ship Island Pass from the Gulf of Mexico. Vaulted casemated batteries were in the first tier and guns en barbette were to be mounted on the second-tier ramparts of the circular front. Toward the rear in the gorge area, the design had flanking structures. In a demibastion on each side of the gorge wall were flanking gun embrasures that guarded the fort's sallyport. In the gorge curtain was the sallyport, drawbridge, and four musket loopholes.

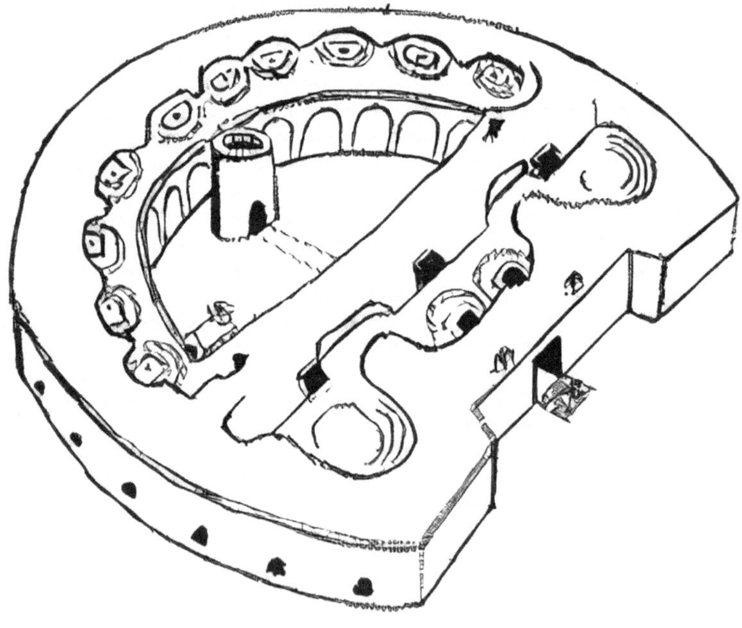

Isometric drawing of Fort Massachusetts Dale Manuel Collection

Fort Massachusetts from the landing on the Mississippi Sound side
Photo by John Weaver

On the fort's parade is a circular stair tower and a hot-shot furnace. The stair tower was once capped by a covering wooden cupola. All the casemates opened out to the parade for ventilation of gun smoke. On the parade's gorge are the entrances to two guard rooms, the sallyport, and two magazines in the first tier. Also on the parade's gorge are two built-in circular stair towers that led to the gorge's second-tier batteries and covered magazines. Earth embankments cover four service magazines and serve as traverses to help block incoming shells, protecting the four gorge barbette gun emplacements. Each end of the embankment contained a magazine. In the center of the embankment, a gallery led to two more magazines.

The parade area of the fort is lined with vaulted casemates up front with the sally port and guard rooms on the left. On the second tier are gun emplacements.
Photo by John Weaver

The fort's circular sides and land defense gorge area Photo by John Weaver

When Mississippi seceded from the Union in January, 1861, state militia seized the unfinished fort. The fort's walls were only 6-8 feet high when the constructing Federal army engineers were forced off the island. In July, Confederate troops mounted several guns at the fort. On July 9, there was a 20-minute exchange of gun fire with the Union blockade ship *Massachusetts* that had come close to the island. These would be the fort's only shots fired in anger. The fort would later receive its unofficial name from this ship; the ship's guns were moved to the island fort. The Confederates abandoned the island in September when it was clear the South could not hold such an isolated position without a strong navy.

Federal troops occupied the island and used it as a staging area for the invasion of New Orleans in the spring of 1862. More than 18,000 troops were stationed on Ship Island during the New Orleans campaign. Almost 40 wooden buildings were constructed on the island during the war including barracks, a hospital, a mess hall, and a

The 15-inch Rodman gun mounted on the second tier gorge battery. Photo by John Weaver

Stair tower Photo by John Weaver

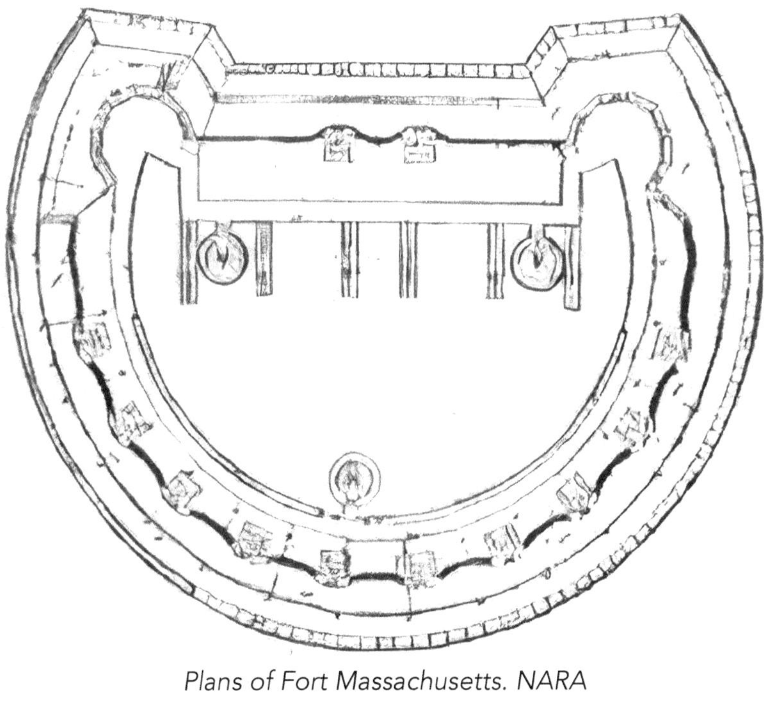

Plans of Fort Massachusetts. NARA

The sallyport Photo by John Weaver

bakery. The island also served as a prisoner-of-war camp and a United States military prison during the war. The military prison operated on the island until 1870. At one time more than 4,000 Confederate prisoners were held on the island. One hundred and fifty-three Confederates died on the island and are buried there.

The Army Corps of Engineers resumed construction of the fort in 1862. There is a clear line of demarcation between the brownish-yellow brick on the bottom and the cherry-red brick on the top portion. This line marks the progress of work before the war. Federal builders could not get local brick during the war and had to import the red brick, as well as slate, granite, and everything else needed to complete the fort from Northern sources. Construction of the fort ended in 1866.

View of the semicircular battery Photo by John Weaver

The interior of the gorge demibastion Dale Manuel Collection

The fort was under caretaker status until 1903 when the last ordnance sergeant left. In 1901, all the guns except the remaining 15-inch Rodman were scrapped. This Rodman gun was one of two that were installed in 1873 and was one of the larger guns made during this period. Only the two experimental 20-inch Rodman guns were larger. The tubes on these smoothbore 15-inch guns weighed over 50,000 pounds and fired a 300-pound projectile about three miles. Two 100-pounder Parrot rifles had been mounted in the center gorge battery. Though the fort was designed to mount 37 guns, only 17 were ever installed. In the casemates of the fort were mounted 10-inch Rodman guns.

Because the island has migrated, the fort that was once 500 feet from the western end of the island is now one mile east of the channel and is on the northern edge of the sound. Ship Island and the fort are administered by the Gulf Islands National Seashore. The fort is one of the best preserved and best maintained examples of these great brick forts in the United States.

The island can only be reached by private boat or by an excursion boat that runs seasonally from Gulfport, Mississippi.

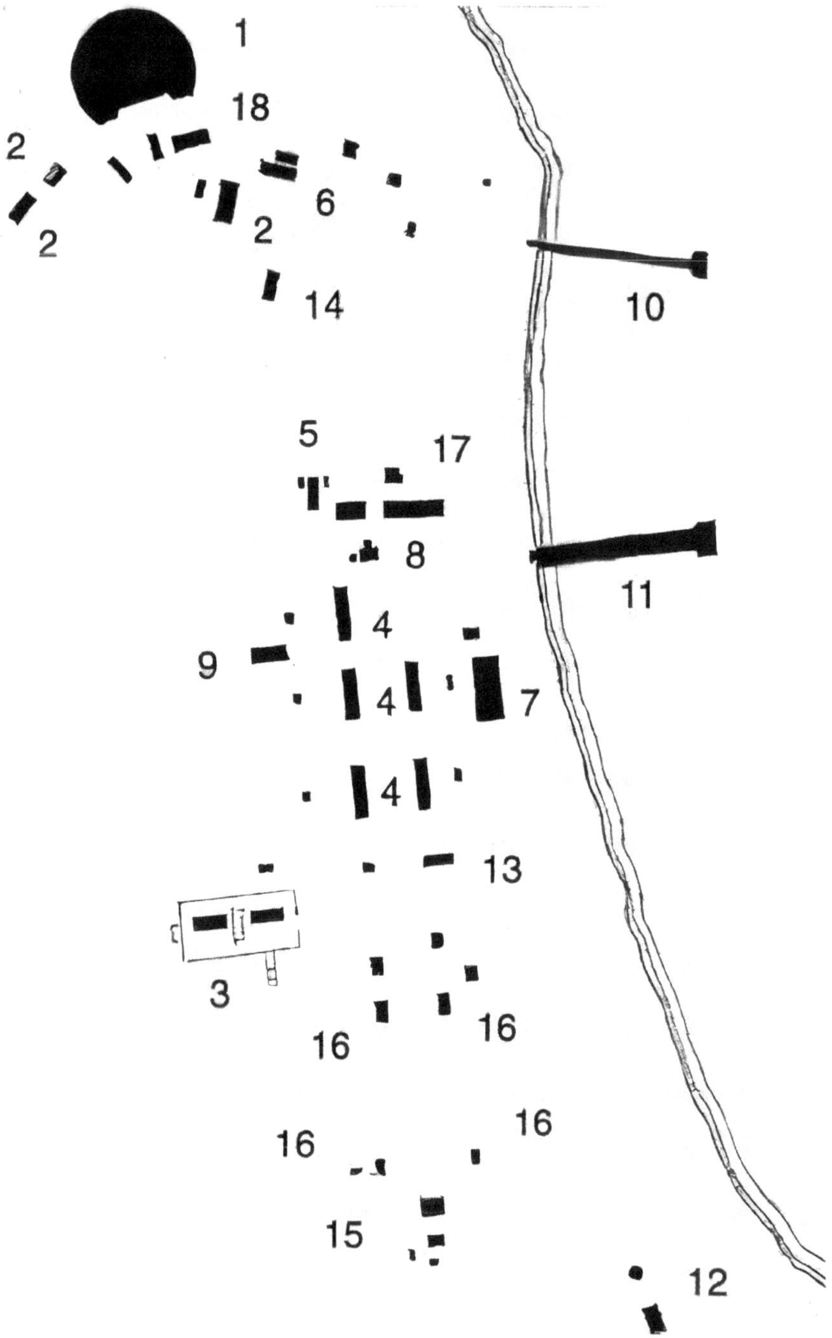

Map of Ship Island during the Civil War. Legend of buildings: 1. Fort; 2. Quarters; 3. Prison Stockade; 4. Barracks; 5. Bakery; 6. Shop; 7. QM Stores; 8. Guard House; 9. Stable; 10. Engineer's Wharf; 11. QM Wharf; 12. Light House; 13. Sutlers Store; 14. Office; 15. Headquarters; 16. Officers' Quarters; 17. Hospital; 18. Store House Dale Manuel Collection

Gorge batteries Mark Berhow Collection

A hot-shot furnace is on a corner of the parade. When heated red hot, shot could be fired at ships where it could start fires or ignite the magazine. Photo by John Weaver

Conclusion

The southern Atlantic and Gulf coasts of the United States were critical for trade with the northern Atlantic coast and trade and goods coming from west of the Appalachian Mountains. Goods were shipped to the Mississippi River, through New Orleans and through the Florida Strait into the Atlantic Ocean. Goods from the Southern states were shipped directly from the ports on the southern Atlantic coast.

From the time that the United States first became a nation, the protection of this coastline was considered vital to the survival of the nation. As a result, masonry – and in some cases earthen – forts were constructed in three consecutive systems. The most expansive and sophisticated of these defense systems, called the Third System, spanned the years 1816 to 1867. Forty-two forts were built during this period to stave off the advances of foreign powers who threatened the sovereignty of the nation. Ironically, these forts never saw action against a foreign power, but became both the catalyst for a rebellion of the southern states and a major platform on which that war was fought. The American Civil War was where these forts were tested and where they proved their mettle.

During the Civil War, advances in iron artillery caused the casemates of these forts to be too small to mount the behemoths necessary to stop steam-powered ironclad ships. Financial woes in the aftermath of the war precluded the reconstruction of these forts and a program of earthen batteries, often in the shadow of, or on the ramparts of, a Third System fort, were constructed to emplace this larger artillery.

By 1886, artillery had been revolutionized by the development of steel cannon with incredible firepower. Masonry forts became obsolete and were replaced by concrete gun batteries with minimal landward defenses. The golden age of forts was over. These long-range guns, supported by mine fields and indirect fire control, took over coastal defense for the next 64 years. By 1950, all fixed fortifications in the United States were considered obsolete and were replaced by missile systems, both land-based and floating, and aircraft.

For more information an American seacoast fortifications:

Seacoast Fortifications of the United States, an Introductory History, by Emanuel R. Lewis, Naval Institute Press, Annapolis, MD 1970, 1992.

Artillerists and Engineers: The Beginnings of American Seacoast Fortifications, 1794-1815 by Arthur P. Wade, CDSG Press, McLean, VA 2011

A Legacy in Brick and Stone: American Coastal Defense Forts of the Third System 1816-1867, By John R. Weaver II, Second Edition, Redoubt Press, 2018.

American Seacoast Defenses, a Reference Guide, 3rd Edition, Mark A. Berhow editor, CDSG Press, McLean, MD 2015

The CDSG Publications *News/Journal* available from the CDSG ePress, cdsg.org/shop/

Appendix: Armament

During the era of the Third System of fortification there were many attempts by the Army to standardize its arsenal of guns and gun carriages. During this time, many developments and improvements in artillery took place. Many of these innovations were in the area of seacoast artillery. Three types of seacoast carriages were designed to fit the special needs of the coastal forts. One type was for casemate emplacements that mounted guns (heavy artillery). Another type mounted guns en barbette on the open terrepleins of these forts. The third type mounted the smaller flank-defense guns.

For flank defense in casemated bastions and counterfire rooms, carronades and 24-pounder howitzers were used. Carronades were short guns of large bore that had

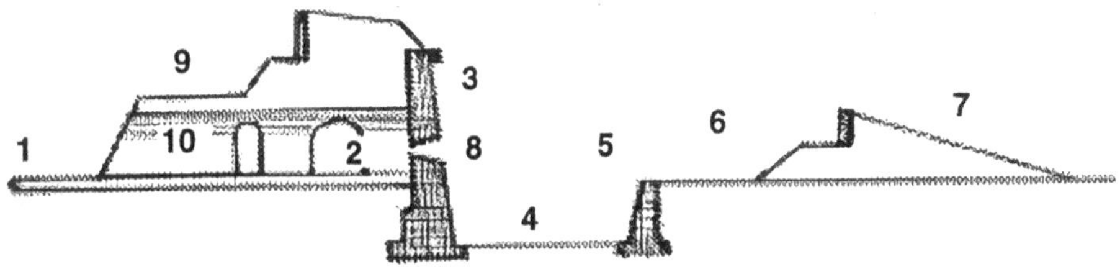

Profile of a fort: 1. Parade; 2. Casemate; 3. Scarp; 4. Ditch; 5. Counterscarp; 6. Covered Way; 7. Glacis; 8. Embrasure; 9. Terreplein; 10. Gallery
Dale Manuel Collection

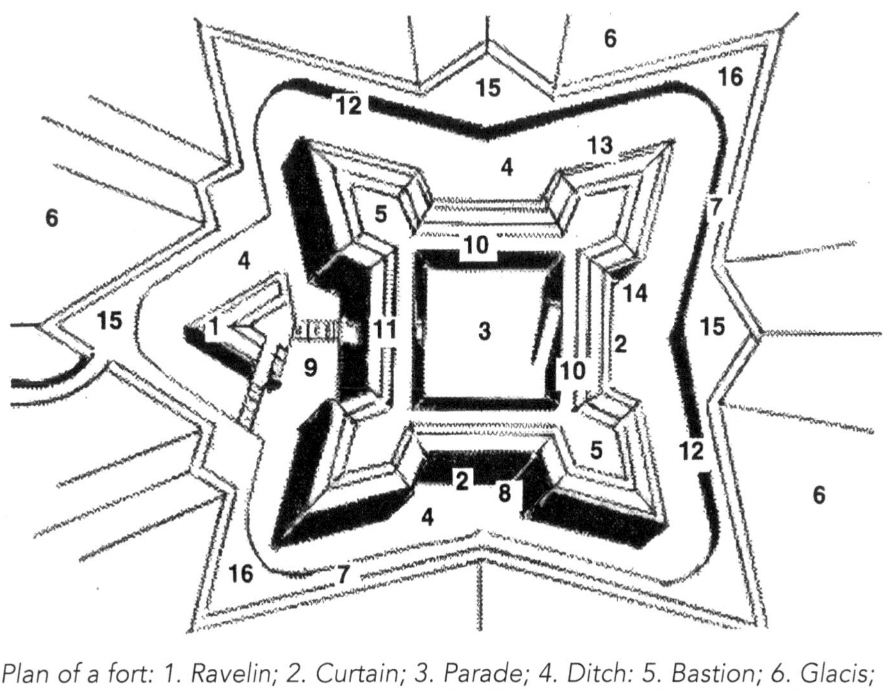

Plan of a fort: 1. Ravelin; 2. Curtain; 3. Parade; 4. Ditch: 5. Bastion; 6. Glacis; 7. Covered way; 8. Curtain angle; 9. Drawbridge; 10. Terreplein; 11. Sallyport; 12. Counterscarp; 13. Face of bastion; 14. Flank of bastion; 15. Re-entering place d'armes; 16. Salient place d'armes; 17. Curtain angle Dale Manuel Collection

been used by the Navy. Both guns' large bores were loaded with grape shot or canister to sweep curtains and ditches with flanking fire. These guns were like very large shotguns firing shot that ranged from pea-size to golf-ball size. Special carriages were designed for these flank guns so they could be fired through the embrasures of flanking works.

One of the first developments was the 1819 model of a 24-pounder smoothbore gun. These guns could throw a 24-pound shot a maximum range of approximately one mile at five degrees elevaion. These iron guns were cast right up to the time of the Civil War. In 1829, the 32-pounder was developed with approximately the same range as the 24-pounder. Later in 1831, 42-pounders were added. These 32-and 42-pounders became very important in seacoast defense. In 1839, new wooden models of car-

The reverse arch was an engineering feature that provided more stability to the foundation of a fort built on wet soil or sand Photo by John Weaver

riages were designed. In 1841, 8- and 10-inch seacoast howitzers were developed that fired shells. In that same year, 10- and 13-inch mortars were developed. A major change occurred in 1844 with the development of two new Columbiads in 8- and 10-inch calibers. The 10-inch fired a 125-lb shot or shell a maximum of three miles, while the 8-inch fired a projectile of 64 lbs the same distance. The new Columbiads

The interior of a wood-lined magazine with stacked barrels of gunpowder Photo by John Weaver

A hot-shot furnace Photo by John Weaver

featured a new carriage with a ratchet device that permitted a maximum elevation of 40 degrees. A normal gun of that period had only a maximum elevation of fifteen degrees. In 1858, the Columbiads were redesigned again, but three years later, a totally new Columbiad was developed that would replace the 8- and 10-inch Columbiads.

This new development was the Rodman method of casting guns. The Rodmans were smoothbore guns with tubes shaped to withstand the stressloads developed when the gun was fired. The most revolutionary feature about the Rodmans was the way they were cast. A normal gun was cast and cooled from the outside in. This method caused stress to be directed outward. Often, during firing, the old guns would burst. The Rodmans were cooled from the inside out, which made them more reliable by causing the stresses in the iron to be compressive. Rodmans were cast in 8-, 10-, 15-inch, and two experimental 20-inch monsters. Rodmans were in service up to the turn of the twentieth century.

Also developed during the Civil War were rifled guns and iron carriages. Elongated projectiles would engage the rifling grooves in the gun. The projectile would spin in flight. This spin and elongation created more accuracy, range, and destructive impact. Parrott rifles of 100-, 200-, and 300-pounder bores were developed. New wrought-iron gun carriages and mortar beds were developed. Innovations in ammunition and new fuses for shells came about.

A 32-pounder on a barbette carriage Photo by John Weaver

A Totten embrasure with shutters Photo by John Weaver

15-inch Rodman gun on a center-pintle carriage Mark Berhow Collection

24-pounder flank howitzer on its special carriage Photo by John Weaver

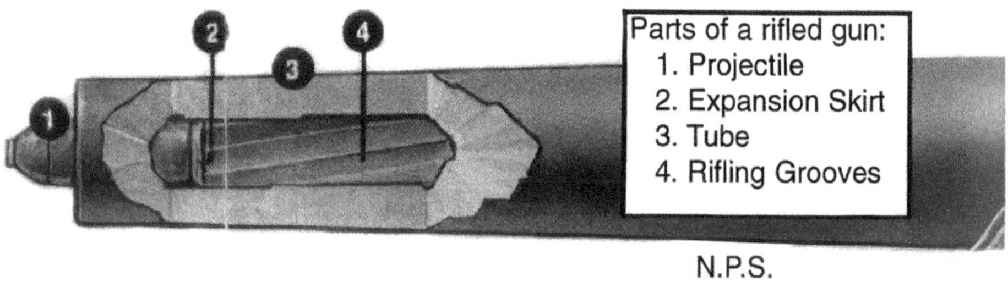

Parts of a rifled gun: 1. Projectile; 2. Expansion skirt; 3. Tube; 4. Rifling Grooves
National Park Service

9-ton, 13-inch mortars could throw a 218-pound shell a maximum high-trajectory range of 4200 yards Photo by John Weaver

Disappearing gun in the down position Photo by John Weaver

6-inch rifle mounted on a disappearing carriage Photo by John Weaver

This 3-inch, rapid-file rifle guarded the minefield against fast torpedo boats and minesweepers John Weaver Collection

The 12-inch mortars of Battery Laidley at Fort DeSoto Park (Mark Berhow)

This submarine mine was anchored in place and connected to the mining casemate by a cable network. With an enemy ship in the mine field, a mine containing 100-200 pounds of explosives could be detonated. Dale Manuel Collection

Gun on a carriage in a casemate. The gun was aimed by rolling the carriage on a traverse arc on the floor. The gun would be rolled forward to be fired through the embrasure. Photo by John Weaver

Appendix: Machinery of the Forts

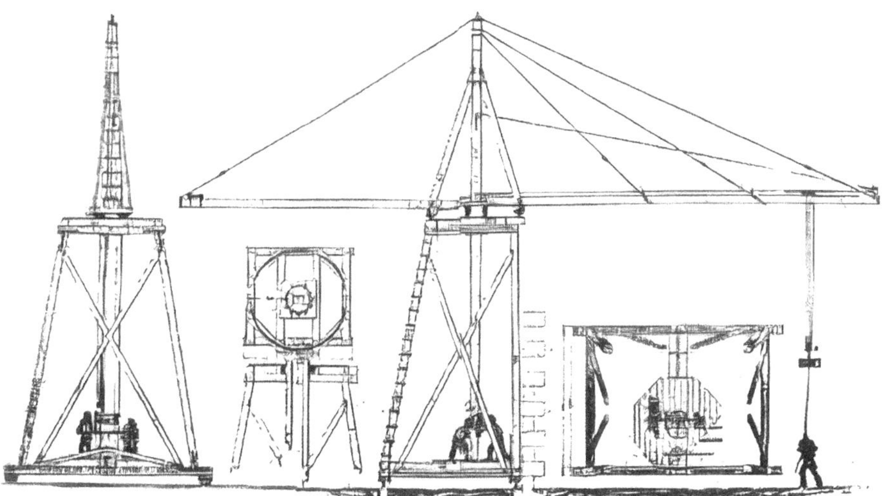

This hand-winched crane could swivel and be rolled on a track Dale Manuel Collection

The Machinery of the Forts

Throughout the Third System most of the projects were plagued by construction delays. Various reasons ranged from delayed or no congressional funding, diseases such as yellow fever epidemics, a shortage of or poor quality materials, settling foundations, labor shortages, storm damage, shortages of trained engineers, etc. In many cases construction took a decade or more. In order to bring the construction to a conclusion, many engineers were willing to use new methods and powered machinery to lessen the work load.

Steam power began to be used more and more. A boiler produced the steam that would turn the engine and a wheel. Belts transferred the rotational torque of the engine wheel to a wheel on the machine. Steam powered circular saws were used to cut wooden piles, timbers, and planks. Steam powered pile drivers were used to construct wharves. Where wet, soft soil or sand existed, piles were driven deep into the ground. A grillage of timbers and planks was laid on top of the piles. On top of the grillage, the foundations of the forts were laid. This system would support the weight of the walls and the artillery atop the ramparts.

A steam powered dredge with a scoop on a long arm was often used to dredge the channels and provide fill. Fill was used to raise the plane of the fort or to fill the parade. In a few cases a dredge was used to excavate the underwater foundations of forts built on shallow shoals. Mechanical cranes were used to load/unload supplies and materials from boats and barges. Cranes lifted construction material to higher tiers. Many used hand cranks, gears, and blocks-and-tackle to raise or lower materials.

In a few cases where forts were built on shallow shoals, a diving bell was used to lay foundation stones underwater. A crane would lower and raise the bell. A hand pump would supply air to the diver in the bell.

A gin was used to lift large guns onto their carriages. A gin was a tripod of large poles with a block-and-tackle arrangement. With a gin, rollers, jacks, and a sling cart, guns weighing more than 25 tons could be moved and mounted. Although these machines may appear to be primitive, they were high-tech for their age. Many were engineered and built on the construction site.

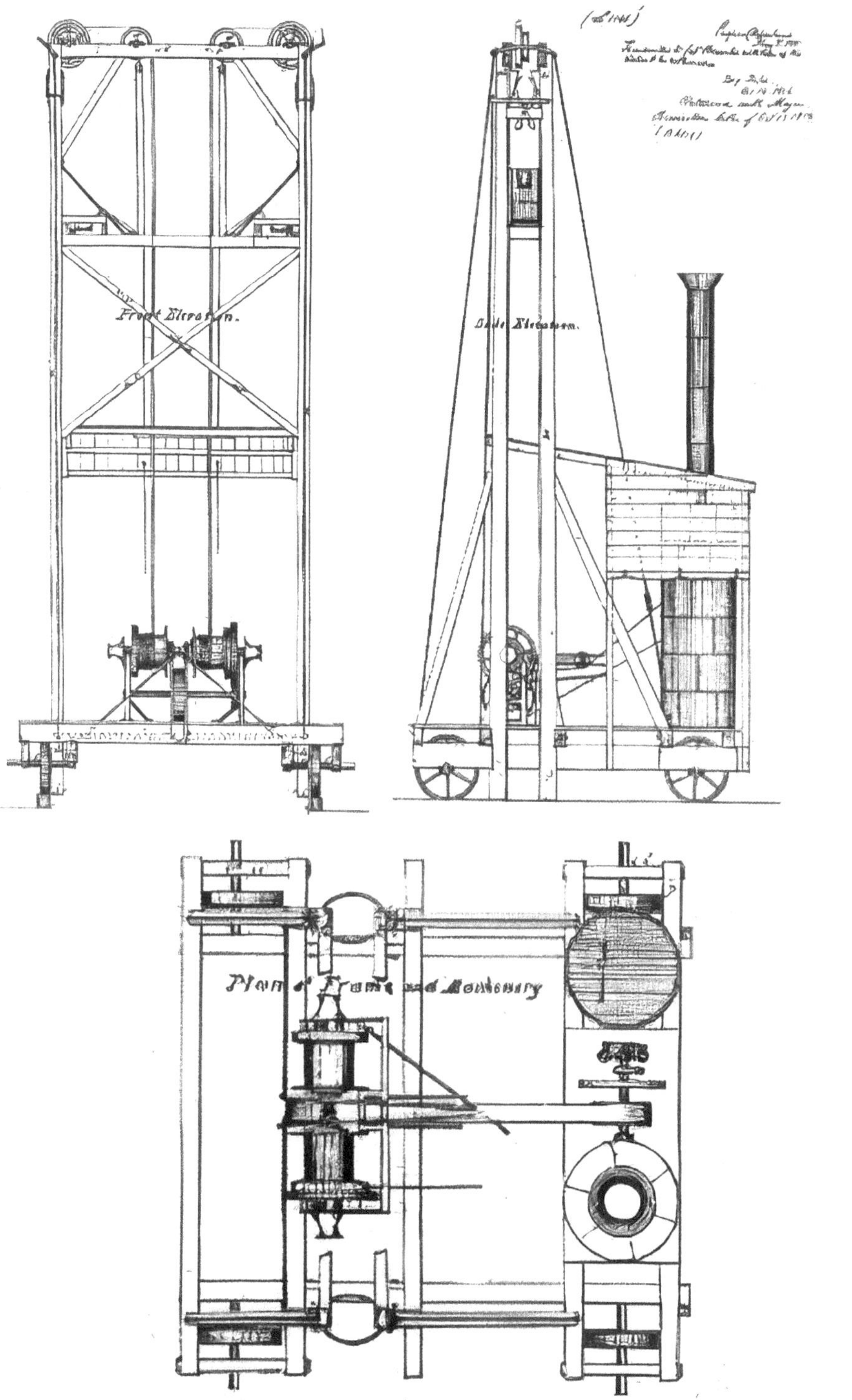

This plan of a steam pile-driver was signed by Robert E. Lee
Dale Manuel Collection

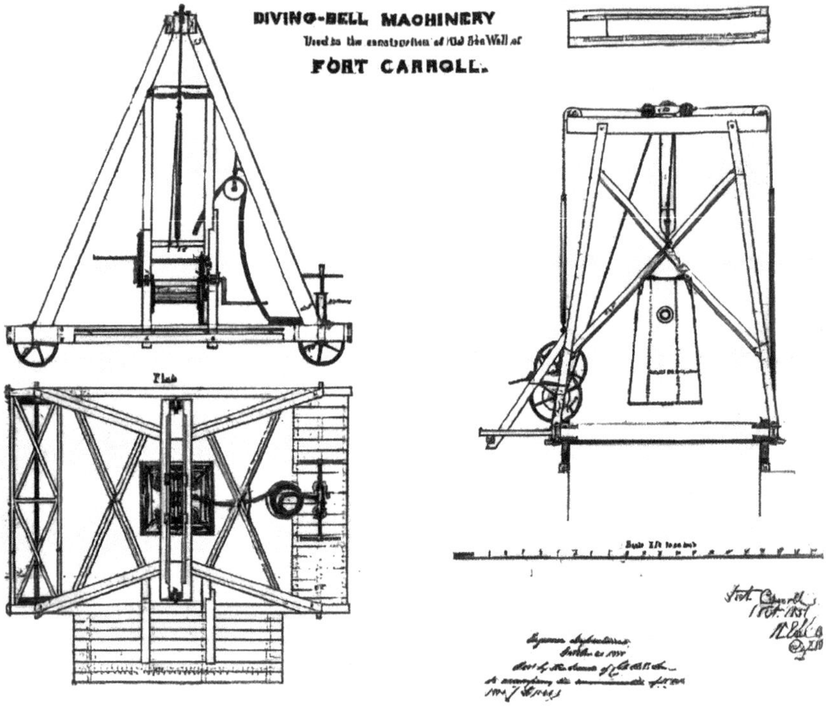
DIVING-BELL MACHINERY
FORT CARROLL.
Plan

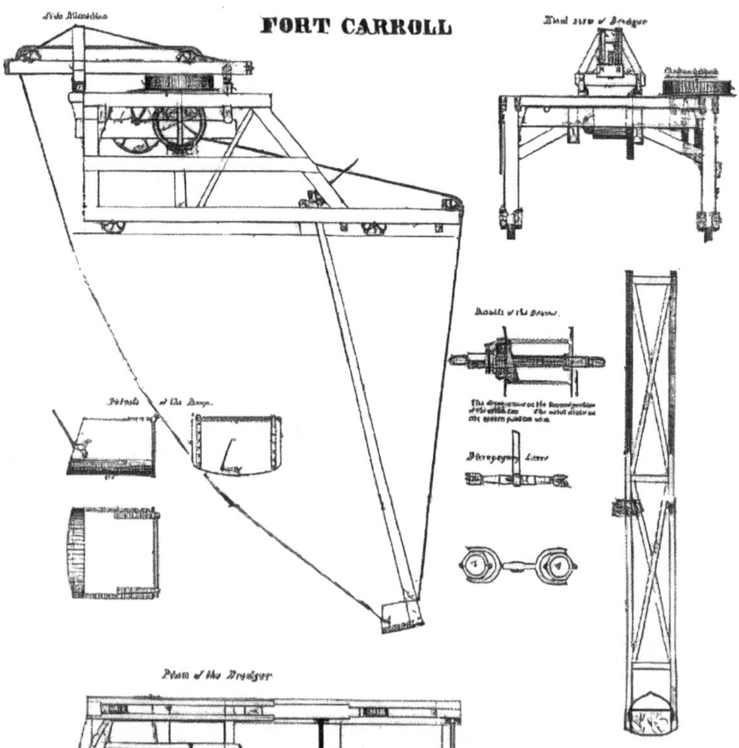
FORT CARROLL

(Previous page top) The drive wheel of this dredge was connected to a steam engine by a belt Dale Manuel Collection.

(Previous page bottom) In this plan of a diving apparatus, signed by Robert E. Lee, the diving bell is lowered and raised by a hand-winched crane. Air is supplied by a hand pump. Lee had plans drawn for every piece of machinery while he was the supervising engineer at Fort Carroll, Maryland. This shows Lee's attention to detail. Dale Manuel Collection

A sling cart moved heavy guns to the emplacements Dale Manuel Collection

Gun crew mounting a large gun with the use of a gin. A gin was a tripod of poles with a block-and-tackle arrangement. Dale Manuel Collection

Glossary of Terms

Advanced Work: A work in advance of a fort, generally within range of the cannon of the day.

Banquette (bahn KET): The raised earthen or paved platform behind the parapet on which riflemen can stand when firing.

Banquette Slope: The earthen or paved slope leading from the banquet to the covered way or terraplein. Riflemen could use this slope for reloading their weapon while shielded from direct fire by the parapet.

Barbette (bar BET): The platforms behind the parapet on which guns are mounted, such that they can fire over the wall. Guns are said to be mounted en barbette when they are mounted on these platforms.

Bastion (BAST yun): A projection from the trace of the fort, usually at a salient, which allows flanking fire along the wall. A full bastion has two faces and two flanks; a demibastion has one face and one flank.

Battery: 1. A structure, generally smaller than a fort and usually open-backed, designed to mount guns. A battery is generally constructed at a location that does not require a fort. 2. A group of gun emplacements, either within a fort or in the outworks of a fort. 3. A group of guns commanded by a single officer.

Bombproof: A structure protected from indirect or plunging fire; a place of refuge during a siege.

Breach: An opening created in a wall, most often during a siege. A breach is generally created by cannon fire.

Carnot Wall (CAR noh wall): A detached scarp; a scarp wall which is separated from the ramparts of a fort, usually by a chemin de ronde. A Carnot wall was an infantry-defense device that was considered expendable in a siege.

Casemate (CASE mate): A vaulted, bombproof room of masonry construction, sometimes called a gunroom. It provides overhead protection to gunners, and allows tiers of guns to be stacked. Casemates may also provide firing positions for small arms, and may be used for quarters, kitchens, storage, and other sundry functions.

Center-Pintle Carriage: A gun carriage with the pintle mounted at or near the center of the carriage, allowing a 360° rotation of the gun.

Chemin de Ronde (SHEH min deh RON dah): A pathway around the interior of the scarp wall of a detached-scarp fort, serving as a firing platform for riflemen.

Citadel (SIT ah dell): A stronghold inside the scarp of a fort which serves as a defensive barracks and a last line of defense.

Counterscarp: The wall across the ditch from the scarp. The counterscarp served as a revetment as well as an obstruction to entry of the ditch by making it a significant drop to the level of the ditch.

Counterscarp Gallery: Casemated areas behind the counterscarp wall, allowing a cross-fire into the ditch.

Coup de Main (KOO dah MIN): An assault of a fortification by storm, rather than by siege.

Coverface: An earthen, earth-and-masonry, or masonry outwork designed to protect the scarp of the fort from direct artillery fire.

Covert (KUV ert) **Way,** or **Covered Way:** The area above the counterscarp, covered by the guns of the fort but protected from view by an attacker by the parapet.

Crenel: An open-topped embrasure at the top of a wall. A crenel is defined by two adjacent merlons.

Crochets (croe SHAYS): Angled or curved paths around a traverse that spans a covert way, generally wide enough for only one soldier abreast to pass.

Curtain: The portion of the scarp between bastions.

Curtain Angle: The angle formed by the curtain and the flank of a bastion. Also called the flank angle.

Demibastion (DEM mee BAST yun): A half-bastion; a bastion with only one face and one flank.

Detached Bastion: A bastion separated from the scarp wall, often by a ditch or chemin-de-ronde. The detached bastions in the Third System were connected to the main work by a gallery.

Detached Scarp: A masonry scarp in advance of an earthen rampart, usually separated by a chemin-de-ronde.

Detached Work: An independent work beyond the outworks of a fort, generally located outside of cannon range from the main work. The "Martello Towers" at Key West would be detached works to Fort Taylor.

Ditch: A low area, either wet or dry, which inhibits the passage of the enemy to the scarp wall. The ditch is located between the scarp and counterscarp wall. A wet ditch is usually referred to as a moat.

Drawbridge: A bridge, generally over the ditch, that controls entrance to the work and can be raised or pivoted to deny access.

Embrasure (em BRAY Shure): An opening in the scarp, the counterscarp wall, or the parapet which allows a cannon to fire through it. An open-topped embrasure, rare in the Third System, is also called a crenel.

En Barbette: The practice of mounting cannon on a platform (barbette) such that they fire over a wall rather than through an opening (embrasure) in that wall.

En Casemate: The practice of mounting cannon in a casemate such that they can fire through an embrasure.

Enceinte (awn SAHNT): The body or mass of a fort, inside the ditch.

Enfilade (en fill ODD) or **Enfilading Fire**: Fire from the side. Flanking fire

Exterior Slope: The slope between the cordon and the superior slope. The exterior slope is usually the steepest slope on the exterior of the parapet. It is usually earthen, designed to absorb artillery fire.

Face: The side or sides of the fort which open onto the water, where the primary coastal guns will be located. On a bastion, the face is the wall of the bastion which is nearly parallel to the curtain.

Flank, or **Cheek**: The side. On a bastion, the flank is the portion which is most nearly perpendicular to the curtain of the fort. The flank of a bastion provided the primary fire along the curtain.

Flank Angle: The angle formed by the curtain and flank of a bastion or demibastion. Also called the curtain angle.

Flanking fire: Fire from the side. Enfilading fire.

Front: The side of a fort. On an unbastioned fort, the front is measured from comer to corner. On a bastioned fort, it is measured from the salient of a bastion to the salient of the opposing bastion, including the curtain wall between the bastions.

Front Pintle or **Fore Pintle Carriage**: A carriage where the pintle is attached at or near the front, giving a limited arc of traverse. When firing through an embrasure, the pintle is at the throat of the embrasure and attaches to the carriage with a tongue – a flat piece of metal that passes through an opening below the embrasure. When mounted en barbette, the pintle is at the front of the carriage.

Gallery: A passageway, generally masonry or masonry revetted.

Glacis (glah SEE): The gentle slope beyond the outworks, cleared of all obstructions, which an attacker must traverse to reach the fort.

Gorge: The side or sides of the fort away from the water, where land defense is the primary concern. On a bastion, the gorge is the narrow portion of the bastion toward the parade, also called the throat of the bastion.

Hot-Shot Furnace or **Shot Furnace**: A furnace for heating shot. A masonry structure, it consisted of a fire chamber below rails on which the shot would travel. A loading mechanism sat at one end and an unloading mechanism at the other. Hot shot was used to set fire to wooden ships, sails, and rigging.

In Battery: A gun in firing position, i.e., at the front of the lower carriage with its muzzle at the throat of the embrasure.

Interior Crest: The highest point of the parapet. The top of the superior slope.

Interior Slope: The vertical or near-vertical slope from the crest of the parapet to the banquette, or the terraplein if there is no banquette. This slope is generally masonry or wood.

Investment: The surrounding of a work in preparation for a siege, preventing resupply or reinforcement of the work. The time period of a siege is generally counted as beginning when the work is invested.

Land Front: The front of a fort facing the land, where a siege would take place. In the Third System, the land front was generally designed in the style of Vauban with protection of the masonry and levels of defense.

Loophole: A narrow opening in a wall which allows a rifle or musket to fire through it. Also called a rifle slit.

Magazine: The place for storage of powder inside a fort. The main (or storage) magazine would store the bulk of the powder, and day-use (or service) magazines would be secondary storage depots. Magazines were carefully designed to prevent sparks and to provide a dry atmosphere for the powder.

Martello Tower: A round tower of three or more floors mounting heavy artillery en barbette on the top level (generally one to three cannon) and infantry-defense weapons (howitzers, carronades, and small arms) on the lower levels. Based on a design in Corsica which successfully defended a harbor, the entrance was always on the second floor. This nomenclature was extended to encompass any tower-like small fort, whether round or rectangular and with or without outworks.

Merlon: A rectangular projection from the top of a wall. Crenels and merlons alternate on a crenellated wall.

Outworks: Works located within or beyond the ditch, i.e., works outside the enceinte of the fort.

Pan Coupe: A feature of an unbastioned fort, consisting of a wall that connects two curtains. It follows the trace of the gorge of a bastion in a bastioned fort, replacing a sharp salient with two broader salients.

Parade: The open area in the center of a fort, usually used for drilling troops, for barracks area, etc.

Parapet: The low wall along the top of the rampart, generally masonry or masonry-revetted earth, which protects the artillery and the banquette. The highest point of the rampart.

Pintle: The pin around which a gun carriage rotates. A front-pintel carriage for a gun mounted en casemate generally has the pintle at the narrowest part (throat) of the embrasure. A center-pintel carriage, sometimes used on a barbette emplacement, allows a full 360-degree traverse of the gun.

Place d'Armes (PLAHSS deh ARMS): A gathering place for counterattacking forces. A place d'armes can be within a fort or a part of the outworks of a fort.

Polygonal Fort: A fort with the trace of a polygon, as opposed to a bastioned fort or a fort with a rounded trace.

Portcullis (por cue LEE): A wooden or iron gate which bars access but not sight, often on the inner side of a sally port. A portcullis generally opened vertically.

Postern (POST ern): The "back door" of a fort. Posterns are secondary entrances leading through the ramparts and either into or over the ditch. A postern often connect to a ravelin, exterior batteries, or another outwork.

Rampart (RAM part): The full thickness of the walls of a fort, including the casemates.

Ravelin (RAVeh lin): A V-shaped outwork. Ravelins were placed in advance of either a curtain or the gorge of a fort, giving protection to the scarp and/or providing additional batteries.

Redan (reh DAHN): A small, V-shaped, open back outwork placed in advance of either a curtain or the gorge of a fort, protecting the scarp and providing additional batteries.

Redoubt (reh DOOT): Technically, an unbastioned fort some distance from the main fort which is designed to form the first line of defense and provide early warning of an attack. In the Third System, redoubts sometimes had demibastions and the distance to the main fort varied greatly.

Reentering: An angle pointing toward the fort; an angle of greater than 90 degrees, such as where a bastion meets a curtain.

Revetment: A depressed area sustained by masonry or wood, with the surrounding earth providing protection.

Salient: An angle pointing away from the fort; an angle of less than 90 degrees, the "point."

Sally Port: The main entrance or entrances to a fort, often including layers of defense against an invader. The sally port is generally located in the gorge wall.

Scarp, or **Scarp Wall:** The perimeter wall of the fort.

Seacoast Front: The front of a fort facing the water, designed to control passage and/or dominate a harbor.

Shell: A projectile fired from a cannon or mortar that has a hollow interior, that is filled with explosives.

Shot: A solid iron projectile fired from a cannon. Shot had far more momentum than shell, but did its damage through momentum only.

Shoulder: The place on a bastion or demibastions where the face and the flank meet.

Siege: The protracted taking of a fort, beginning with investment and continuing through the breaching and ultimate reduction of the work.

Sole: The bottom of an embrasure or loophole.

Superior Slope: The slope that reaches the crest of the parapet. The superior slope is usually the shallowest slope of a parapet.

Terraplein (TARE a pleh): The area atop the ramparts behind the banquette and parapet, supporting the barbettes.

Throat: The narrowest part. On a bastion, the narrow portion between the flanks, nearest the parade. On an embrasure, the narrowest portion of the opening.

Totten embrasure: An embrasure reinforced with iron, developed by J. G. Totten. These embrasures provided a smaller opening and a wider angle of traverse of the cannons than European embrasures. Totten embrasures usually, but not always, utilized Totten Shutters.

Totten shutters: Iron doors which close an embrasure between cannon firings. This intricate system was developed by J. G. Totten.

Tower Bastion: A bastion of reduced size, often of significant height.

Trace: The line defining the exterior of the enceinte, or body, of a fort. The trace is generally referred to by the geometric figure that defines it.

Traverse (trah VERSE): 1. An earthen or masonry-revetted earthen mound which breaks the open area of the terraplein, separating men and guns to eliminate enfilading fire and minimizing damage from exploding shells. 2. The arc of stone on which the rear wheels of guns travel. The individual stones are generally referred to as traverse blocks. The iron rail that the wheels travel on is generally referred to as a traverse rail.

Water Battery: A water battery is a group of cannon facing seaward and firing as a unit. A battery was generally commanded by a single officer. Use of the term water battery often implies a separate, detached work housing a group of cannon.

Work: A fortification. This is the broad term that applies to forts, fortresses, towers, redoubts, batteries, etc.

Articles by Dale Manuel published by the Coast Defense Study Group

- Bombardment of Fort St. Philip *Coast Defense Journal* Volume 20, Issue 3, August 2006 - p. 79
- Caponiers of the Third System *Coast Defense Journal* Volume 21, Issue 2, May 2007 - p. 4
- Carnot's Wall, or the Detached Scarp in America *Coast Defense Journal* Volume 15, Issue 3, August 2001- p. 76
- Characteristics of the Second System *Coast Defense Journal* Volume 15, Issue 1, February 2001- p. 84
- Characteristics of the Third System of American Seacoast Defenses *CDSG Journal* Volume 14, Issue 3, August 2000 - p. 46
- Citadels of the Third System of American Seacoast Defense *Coast Defense Journal* Volume 15, Issue 4, November 2001 - p. 57
- Col. Edward Harcourt, Chief Engineer, Army of the Republic of Texas *Coast Defense Journal* Volume 19, Issue 4, November 2005 - p. 51
- Drawings of Guillaume Tell Poussin CD Journal Volume 16, Issue 1, February 2002 - p. 82
- Fort Fremont and the Defenses of Port Royal Sound, South Carolina *Coast Defense Journal* Volume 16, Issue 4, November 2002 - p. 29
- Fort Livingston: The Lonely Sentinel of Barataria *CDSG Journa*l Volume 14, Issue 4, November 2000 - p. 58
- Fort Maurepas, Vieux Biloxi *Coast Defense Journal* Volume 20, Issue 3, August 2006 - p. 31
- Fort Petite Coquilles, Louisiana *Coast Defense Journal* Volume 17, Issue 2, May 2003 - p. 85
- Fort St. John, Louisiana *Coast Defense Journal* l Volume 16, Issue 2, May 2002 - p. 43
- Fort St. Leon and the Fortifications at English Turn, Louisiana *Coast Defense Journal* Volume 17, Issue 3, August 2003 - p. 62
- Fortifications at Balize, Louisiana *Coast Defense Journal* Volume 17, Issue 3, August 2003 - p. 72
- Fortifications of Pelican Spit, Galveston Harbor *Coast Defense Journal* Volume 21, Issue 1, February 2007 - p. 21

- Forts Hampton, Winyaw, & Marion: 2nd-System Forts of the Carolinas *Coast Defense Journal* Volume 17, Issue 4, November 2003 - p. 43
- Galveston Forts and the Storm of the Century *Coast Defense Journal* Volume 15, Issue 2, May 2001- p. 4
- Land Defenses of New York City During the War of 1812 *Coast Defense Journal* Volume 36, Issue 3, Summer 2022 - p. 10
- Major C.E. Blunt's Other Civil War Forts *Coast Defense Journal* Volume 33, Issue 3, Summer 2019 - p. 49
- Massachusetts North Shore Civil War Forts *Coast Defense Journal* Volume 33, Issue 3, Summer 2019 - p. 27
- Robert E. Lee and the Construction of Fort Carroll, Maryland *Coast Defense Journal* Volume 18, Issue 3, August 2004 - p. 40
- Some Memories of the Coast Artillery Corps Citizens' Military Training Camps *Coast Defense Journal* Volume 18, Issue 1, February 2004 - p. 55
- Spanish-American War Batteries at Sabine Pass *Coast Defense Journal* Volume 16, Issue 3, August 2002 - p. 59
- Tours-Modèles in America *Coast Defense Journal* Volume 18, Issue 4, November 2004 - p. 65
- Vanished Fort McRee - *Coast Defense Journal* Volume 33, Issue 4, Fall 2019 - p. 53

THE COAST DEFENSE STUDY GROUP, INC.

CDSG.ORG

The Coast Defense Study Group, Inc. (CDSG) is a tax-exempt corporation dedicated to the study of seacoast fortifications. The purposes of the CDSG include educational research and documentation, preservation and interpretation of historic sites, and assistance to other organizations dedicated to the preservation and interpretation of coast defense sites. Membership is open to any person or organization interested in the study or history of coast defenses and fortifications. Membership in the CDSG will allow you to attend annual conferences, special tours, and receive quarterly newsletter and journal. To find our more about the CDSG, please visit the CDSG website at **cdsg.org.**

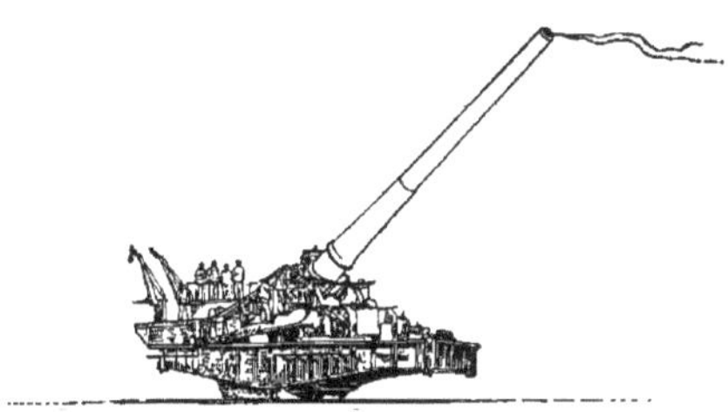

The CDSG Press

Coast Defense Study Group Press is a division of Coast Defense Study Group which publishes books of historical interest, especially concerning seacoast fortifications. The CDSG Press also offers an ever-expanding number of key reprints reports and manuals in electronic PDF format on compact disks. To order these books and other **CDSG Press** publications, please access the **CDSG Press** pages on the **CDSG web site** at **cdsg.org.**

CDSG Press is interested in new titles, especially those dealing with fortifications, please contact Terry McGovern at 703/538-5403 or at tcmcgovern@att. net if you have a title that you are seeking to have published. Visit www.cdsg.org/press.

Under the CDSG Press label, our organization has published:

Notes on Seacoast Fortification Construction by Col. Eben E. Winslow, 1920, 428 pp. 1994 reprint HC with bound drawings
Seacoast Artillery Weapons Technical Manual (TM) 9-210 by U.S. War Dept. 1944, 202 pp. 1995 reprint PB
The Service of Coast Artillery by F. Hines & F. Ward, 1910, 736 pp. 1997 reprint HC
Permanent Fortifications & Sea-Coast Defenses by U.S. Congress, 1862, 544 pp. 1998 reprint HC
American Coast Artillery Material Ordnance Dept. Doc#2042 by U.S. War Dept., 1922, 528 pp., 2001 Reprint HC
American Seacoast Defenses: A Reference Guide (3rd Edition) by Mark A. Berhow, 2015, 732 pp. HC
The Endicott & Taft Board Reports, reprint of original reports of 1886 and 1905 by U.S. Congress, 525 pp. 2007 HC
Artillerists and Engineers: The Beginnings of US Fortifications 1794-1815 by Arthur P. Wade, 226 pp. 2011 PB
World War II Harbor Defenses of San Diego by Commander (Ret.) H.R. Everett, U.S. Navy, 226 pp. 2020 HC
American Seacast Forts - A Directory 1890-1950 , in 4 volumes, by T. McGovern, M. Berhow and G. Williford (2025) Vol. 1 North Atlantic Coast 376 pp, Vol. 2 South Atlantic and Gulf Coasts 326 pp., Vol. 3 Pacific Coast 264 pp., Vol. 4 Alska, Hawaii, the Phillipines, Panama, Nova Scotia, and the Carribean 336 pp. HCs

The CDSG Presss offers these Hole in the Head Press Books

*Artillery at the Golden Gat*e by Brian B. Chin (Hole-in-the-Head Press), 176 pp. 2019 PB
Fort Baker Through the Years by Kristin L Baron and John A. Martini (Hole-in-the-Head Press), 99 pp. 2013 PB
Rings of Supersonic Steel (3rd Edition) by Mark Morgan and Mark Berhow (Hole-in-the-Head Press), 358 pp. 2010 PB
The Last Missile Site by Stephen Hailer and John A. Martini (Hole-in-the-Head Press), 158 pp. 2010 PB
To Defend and Deter by John Lonnquest and David Winker (Hole-in-the-Head Press), 432 pp. 2014 PB

The CDSG Press
1700 Oak Lane
McLean, VA 22101-3322 USA

www.ingramcontent.com/pod-product-compliance
Lightning Source LLC
LaVergne TN
LVHW050909210926
848482LV00012B/378
9798994798904